Springer Aerospace Technology

The series explores the technology and the science related to the aircraft and spacecraft including concept, design, assembly, control and maintenance. The topics cover aircraft, missiles, space vehicles, aircraft engines and propulsion units. The volumes of the series present the fundamentals, the applications and the advances in all the fields related to aerospace engineering, including:

- structural analysis,
- aerodynamics,
- aeroelasticity,
- aeroacoustics,
- flight mechanics and dynamics
- orbital maneuvers,
- avionics,
- systems design,
- materials technology,
- launch technology,
- payload and satellite technology,
- space industry, medicine and biology.

The series' scope includes monographs, professional books, advanced textbooks, as well as selected contributions from specialized conferences and workshops.

The volumes of the series are single-blind peer-reviewed.

To submit a proposal or request further information, please contact: Mr. Pierpaolo Riva at pierpaolo.riva@springer.com (Europe and Americas) Mr. Mengchu Huang at mengchu.huang@springer.com (China)

The series is indexed in Scopus and Compendex

More information about this series at http://www.springer.com/series/8613

Valery Dmitryevich Sharov ·
Vadim Vadimovich Vorobyov ·
Dmitry Alexandrovich Zatuchny

Probabilistic-Statistical Methods for Risk Assessment in Civil Aviation

Springer

Valery Dmitryevich Sharov
Moscow, Russia

Vadim Vadimovich Vorobyov
Moscow, Russia

Dmitry Alexandrovich Zatuchny
Moscow, Russia

Translated by Kudriashova Anna

ISSN 1869-1730 ISSN 1869-1749 (electronic)
Springer Aerospace Technology
ISBN 978-981-16-0091-3 ISBN 978-981-16-0092-0 (eBook)
https://doi.org/10.1007/978-981-16-0092-0

This Springer imprint is published by the registered company Springer Nature Singapore Pte Ltd.
The registered company address is: 152 Beach Road, #21-01/04 Gateway East, Singapore 189721, Singapore

Preface

Despite a specific improvement in recent years, the state of Safety (S) in Civil Aviation (CA) of the Russian Federation remains unsatisfactory. If, during regular air transportation, the level of S in CA of the Russian Federation corresponds to the global average, then in general, for commercial CA, it is lower than the global average.

There is a significant lag behind the industry-leading countries. So, for the period 2007–2018, according to calculations based on the data of the Federal Aviation Administration of the Russian Federation and the National Transportation Safety Board (NTSB), USA, the average annual relative number of Accidents (ACCID) per 100 thousand flight hours was 0.538 in the commercial CA of the Russian Federation and 0.324 in the USA. The relative number of disasters in the Russian Federation is 0.259 and in the USA 0.099. The number of deaths per 1 million transported in the Russian Federation is 0.870 and in the USA 0.044.

A current approach to problem-solving of increasing S provides for the development of an Aviation Safety Management System (SMS) for each airline, and the basis of the SMS is the process of managing the risk of air safety.

There is no standard for SMS in the Russian Federation, and airlines develop SMS independently. Under these conditions, in the absence of a standard S risk management methodology and a lack of guidelines for the development and implementation of methods, there is a broad and subjective interpretation of several provisions of the ICAO Safety Management Manual (SMM) and the Russian Air Law. This leads to a conflict of priorities and misallocation of the resources of airlines and, ultimately, does not allow providing a satisfactory level of S in CA of the Russian Federation.

The approach to S management recommended by ICAO is not fundamentally new. It has long been used in the framework of solving the problem of ensuring S, which is the work of Barzilovich E. Y., Bachkalo B. I., Vorobyev V. V., Vorobyev V. G., Guzia A. G., Gipich G. N., Evdokimov V. G., Elisov L. N., Zubkov B. V., Kovalenko G. V., Kuklev E. A., Lebedev A. M., Makarov V. P., Matveev G. N., Livshits G. L., Lushkin A. M., Neymark M. I., Plotnikov N. I., Rukhlinsky V. M., Sakacha R. V., Stolyarov N. A., Shapkin V. S., and others.

It should be noted the development of the State flight safety center, the State research Institute of civil aviation, the Research institute of air navigation, the

Moscow State technical university of civil aviation, the St. Petersburg State university of civil aviation and the Ulyanovsk Higher aviation school of civil aviation.

Foreign organizations include the National Aeronautics and Space Administration (NASA), the USA; the Commercial Aviation Safety Team (CAST), the USA; the Royal Netherlands Aerospace Centre (NLR), the Netherlands; and the Airline Risk Management System (ARMS) at EASA, S departments of Boeing and Airbus corporations.

On the other hand, there are serious theoretical studies and practical results on the assessment of technogenic risks in hazardous industries in the works of the Institute of Control Sciences Academician V.A Trapeznikov RAS, Mechanical Engineering Research Institute of the RAS, as well as those of authors such as Alexandrovsky L. N., Aronova I. Z., Brushlinsky N. N., Belova P. G., Viktorova V. S., Vishnyakov Y. D., Vladimirov V. A., Vorobiev Y. L., Kovalevich O. M., Kryukov S. P., Kulba V. V., Makhutov N. A., Malinetskiy G. E., Novikov D. A., Orlov A. I., Ryabinin I. A., Raikov A. N., Spesivtsev A. V., Shvyryaev Y. V., and other scientists.

Foreign authors made a significant contribution to solving general and applied safety and decision-making problems: D. Dubois, A. Prad, T. Saati, D. Reason, P. Hudson, N. Leveson, E. Henley, H. Kumamoto, and others.

The principles and methods of risk management are reflected in several GOST RFs and the International Organization for Standardization ISO.

However, methodological materials for managing S risk are not enough. There is a reassessment of the "ICAO matrix" as the only risk assessment method to the detriment of others used in similar areas. The recommendations of ICAO and other organizations are not always specific. For example, when the ARMS group recommends using the "Event Risk Classifier" (ERC) or in the Russian interpretation, "the Deviation and Event Risk Coefficient" (DERC), it does not offer a specific method for assessing and monitoring it.

This book is devoted to the analysis of the risks of an airline under various conditions of natural and artificial origin, taking into account the existing regulatory framework and the construction of risk assessment and management models.

Moscow, Russia

Valery Dmitryevich Sharov
Vadim Vadimovich Vorobyov
Dmitry Alexandrovich Zatuchny

Contents

Abbreviations

AC of RF	Air code of the Russian Federation
ACCID	Accident
ACFT	Aircraft
AD	Aerodrome
AE	Aviation Event
AIP	Aeronautical Information Publication
AL	Airline
ARMS	Airline Risk Management Solution Group
ARMS	Automated Risk Management System
ASFPA	Automated System of Forecasting and Prevention of Accidents
ATIS	Automatic Terminal Information Service
ATLCH	Aircraft Takeoff and Landing Characteristics
ATS	Air Transport System
BPHF	Basic Probability of a Hazard Factor
CA	Civil Aviation
CAP	Civil Aviation Publication (UKCAA)
CAR	Canadian Aviation Rules
CAST	Commercial Aviation Safety Team
CAT	Clear Air Turbulence
CATS	Casual Model for Air Transport Safety
CFIT	Controlled Flight Into Terrain
CNE	Consumer Navigation Equipment
CRFI	Canadian Runway Friction Index
DB	Database
DH	Decision Height
DM	Decision Maker
EASA	European Aviation Safety Agency
ECAST	European Commercial Aviation Safety Team
EDI	Event Damage Indicator
EHF	Enlarged Hazard Factors
EST	Earth Satellite Transmitter
FAA	Federal Aviation Agency

FAR	Federal Aviation Regulations
FAR	Federal Aviation Rules
FATA	Federal Air Transport Agency of Russia
FOM	Flight Operation Manual
GLONASS	Global Navigation Satellite System of Russia
GNSS	Global Navigation Satellite System
GOST	Russian State Standard
GPS	Global Positioning System
GYWSGP	Ground Early Warning Systems for Ground Proximity
HF	Hazard Factor
IATA	International Air Transport Association
ICAO	International Civil Aviation Organization
IE	Initiating Event
INS	Inertial Navigation System
IS	Interference Source
LOC-I	Loss of Control in Flight
LV	Linguistic Variable
MD	Management Decision
METAR	Meteorological Aerodrome Report
MF	Membership Function
MS	Maintenance Service
MS	Meteorological Support
NOTAM	Notice to Airmen
NSC	Navigation Spacecraft
NTSB	National Transport Safety Bureau
PA	Pattern of the Antenna
PIC	Pilot In Command
PNR	Priority Number Risk
PSA	Probabilistic Safety Analysis
QFE	Atmospheric pressure at the aerodrome corresponding to the level of the runway threshold of the landing-operating course
QFF	Pressure at the station, reduced to sea level according to actual weather conditions
QNH	Air pressure at the aerodrome, reduced to sea level according to the standard atmosphere
RA	Risk Assessment
RE	Runway Exertion
RRDE	Risk Ratio for Defections and Events
S	Safety
SAA	Safety of Aviation Activity
SAFA	Safety Assessment of Foreign Aircraft
SARP	Standard and Recommended Practices
SARPs	Standards and Recommended Practices
SID	Standard Instrument Departure
SMM	Safety Management Manual

SMS	System Management of Safety
SRNS	Satellite Radio Navigation System
SSR	State Safety Regulations
STAR	Standard Terminal Approach Route
TAF	Terminal Aerodrome Forecast
TFS	Theory of Fuzzy Set
UAV	Unmanned Aerial Vehicle

Chapter 1
Concept of Risk and Safety: Analysis of Aviation Safety Regulations

1.1 Safety in the Context of National Safety Policy

The problems of Safety management (S) should be considered in the context of the general state safety policy, the formation of which is based on the concepts of sustainable development and acceptable risk [1]. The concepts are based on the recommendations of the UN Conference on Environment and Development, held in Rio de Janeiro in 1992. Sustainable development is understood as ensuring the satisfaction of needs without harming future generations based on 27 principles, the main of which are as follows:

- the right of every person to a healthy and fruitful life;
- socio-economic development should be aimed at improving the quality of life;
- conservation of the environment.

The problems of management and state regulation of safety and risks are solved in Russia based on the Constitution of the Russian Federation, the "Concept of National Safety" (1996) through a system of more than 50 federal laws.

In the Constitution of the Russian Federation, safety issues include articles 20, 37, 41, 42, 44, 55, 71, and 72. The federal laws on the most critical safety issues include the laws: "On safety" (1992), "On Environmental Protection" (1991), "On Radiation Safety" (1995), "On Industrial Safety of Hazardous Production Facilities" (1997), "On Technical Regulation" (2002), and "On Transport Safety" (2007).

About 20 GOST RFs also belong to safety [2]. In this case, the last three laws directly relate to the activities of airlines.

When analyzing these documents, one can see certain contradictions in the definitions of the critical concept of safety.

Released in 1992, the Federal Law "On safety" (and most federal laws) defines the following:

> **Safety** is a state of safety and sustainable development of the individual and society, the state, the environment, and life from internal and external threats.

V. D. Sharov et al., *Probabilistic-Statistical Methods for Risk Assessment in Civil Aviation*, Springer Aerospace Technology, https://doi.org/10.1007/978-981-16-0092-0_1

The Law "On Technical Regulation" (2002) defines another: "Safety of products, processes of production, storage, transportation, and disposal—a condition in which there is no unacceptable risk associated with harm to the life or health of citizens, the property of individuals or legal entities, state or municipal property, environment, life or health of animals and plants."

In our opinion, this contradiction can be resolved as follows [3, 4]. The Law "On safety" formulates this concept in general, conceptual terms, taking into account the main decisions of the UN Conference in Rio de Janeiro, which Russia supported.

The Law "On Technical Regulation" provides a more specific definition of this "condition" for facilities and products with an indication of the principle and tool for its assessment—safety risk management. Following this law, which put forward the requirements for identifying risks, it is necessary to create and develop a new system of general and special technical regulations, as well as national and organization standards. Technical regulations are adopted in order to protect the life and health of people, property, and the environment. They establish, taking into account the degree of risk, the minimum requirements that ensure safety.

The importance of risk management is such that risk management is called one of the leading technologies of our civilization [5].

Risk management is a critical element of any safety management system, including the Aviation Safety Management System (SMS).

Traditionally, air transportation is one of the safest modes of transport, but in reality, the position of air transport, among other modes, depends on the assessment methodology. According to the calculations in the classic work of Henley and Kummamoto, based on data from accidental mortality in the USA in 1969, the probabilities of a fatal premature outcome per year for air and water transport are the same and make up 9×10^{-6}. This is better for the automobile (3×10^{-4}) but worse for the railway (4×10^{-6}). Moreover, the average probability of death from accidental causes is 6×10^{-4} per year. However, this calculation method, based on dividing the number of deaths per year by the US population, cannot fully reflect the danger, because it does not take into account the number of people using each type of transport.

Speaking about the fact that air transport is the safest, they often give the results of calculating the period during which a person flying every day can get into a plane crash. For 25 leading US airlines for the period 1988–2007, this figure was about 16 thousand years, and the probability of dying in a disaster corresponds to daily flights for about 29 thousand years.[1]

The complete safety comparison for different modes of transport is made in the monograph of Plotnikov [6]. According to the results of a study based on Russian statistics from 1992 to 1998, places according to the level of danger of five types of transport were distributed as shown in Table 1.1.

[1] OAGback Aviation Solutions & PlaneCrashInfo.com accident database, 1988–2007.

Table 1.1 Distribution of modes of transport by hazard level (1—the most significant danger) depending on the estimated indicator (according to the monograph of Plotnikov)

Mode of transport	Rate		
	The absolute number of deaths	Number of deaths per 1 million transported	Number of deaths per 100 million passenger-km
Automobile	1	3	1
Rail transport	3	5	5
Air transport	**2**	**1**	**4**
Fluvial transport	4	4	3
Maritime transport	5	2	2

Note that the results can vary greatly depending on the data sources. So, in similar studies [7], data on the number of deaths in EU countries for 2001–2002 per 100 million pass-km in railway and air transport are the same and amount to 0.035, and according to [6], they differ significantly: 0.01 for railway and 0.24 for aviation.

Thus, it is premature to say that air transport is the safest. At the household level, the most important is the indicator of the death toll to the total number of people using this mode of transport, and by this rate, air transport is not the leader. At the same time, safety rates differ significantly for different regions, and, unfortunately, the Russian Federation's refers are the worst.

Following the classification, according to the sources of origin and the affected objects [8], aircraft accidents (ACCID) are related to human-made disasters.

Accidents and disasters are divided into types according to the scale of the countries and territories covered by them, the number of victims, and damage, as shown in Table 1.2 (data from [8]).

Table 1.2 The place of ACCID in the classification of disasters by a scale

	A type	Periodicity	Damage (dollars)	The number of victims	The objects
Zone of ACCID	Planetary			Death	Столкновение с крупным астероидом, ядерная война
	Global	30–40 years	10^9–10^{10}	10^4–2×10^6	Nuclear, military, space rocket
	National	10–15 years	10^8–10^9	10^3–10^5	Nuclear, chemical, military
	Regional	1–5 years	10^7–10^8	10^2–10^4	Chemical, energy, transport
	Local	1–6 month	10^5–10^7	10^1–10^3	Technical
	Object	10^3–10^5	10^0–10^2	10^0–10^2	Technical

Table 1.3 Place of ACCID in the classification of emergencies by severity

Parameter	Emergency class					
	Local	Field	Territorial	Regional	Federal	Cross-border
Number of victims	**<10**	**10–50**	**50–500**	**50–500**	>500	
The number of people with a violation of the standard of living	**<100**	**100–300**	**300–500**	**500–1000**	>1000	
Damage	**<1000**	**1000–5000**	5000–500 000	500,000–5,000,000	>5,000,000	
Zone size	**An object**	**Settlement**	The subject of the RF	Two subjects of the RF	The territory of the RF	Outside of the RF
	Zone of ACCID					

At the same time, ACCID occupies a position among regional, local, or site-specific emergencies. According to the severity of emergencies in the Russian Federation [8], they are classified into six classes (see Table 1.3).

1.2 Comparative Estimation of Safety Situation in the Russian Federation and in the World

1.2.1 Comparison of Safety Levels in the Russian Federation and in the USA

We will assess the condition using official data from the report of the Federal Air Transport Agency "Analysis of the state of the power supply in the Civil Aviation Administration of the Russian Federation in 2018", data from the NTSB [9] and the Bureau of Transportation Statistics of the US (see Tables 1.4 and 1.5).

Table 1.5 shows the results of the calculation of indicators based on S analyzes in the Russian Federation for 2007–2018, performed by the State Service of Civil Aviation, Federal Safety Service and the Federal Aviation Administration, both taking into account these eight events and without them (gray cells).

Based on the calculation results in Tables 1.4 and 1.5, a comparison of the S indicators of the Civil Aviation Administration of the Russian Federation and the USA is performed, and the results are shown in Table 1.6.

Table 1.4 US air transport commercial aviation statistics (according to NTSB [Electronic Resource] Access Mode http://www.ntsb.gov/investigations/data/Pages/aviation_stats.aspx and US Department of Transportation [Electronic Resource]. Access Mode http://www.rita.dot.gov/bts/acts)

Year	Aircraft accidents		Death	Data for all commercial transportation		ACCID for 100 thousand hours		Transported passengers, million	The dead per 1 million transported
	All	Disasters		Hours, thousand	Departures	All	Disasters		
2007	87	14	43	23,047,990	10,734,170	0.38	0.13	726,419	0.06
2008	77	20	69	21,784,892	10,282,575	0.35	0.19	690,249	0.10
2009	73	3	67	20,083,630	9,564,891	0.36	0.03	656,995	0.10
2010	58	6	17	20,348,339	9,467,282	0.29	0.06	670,370	0.03
2011	79	16	41	20,782,219	9,419,064	0.38	0.17	681,353	0.06
2012	63	8	12	20,793,757	9,241,935	0.30	0.09	685,102	0.02
2013	64	10	25	20,772,158	9,268,061	0.31	0.11	690,161	0.04
2014	64	8	20	21,008,105	9,049,491	0.30	0.09	708,401	0.03
2015	66	7	27	21,085,142	8,975,795	0.31	0.08	744,241	0.04
2016	56	7	19	21,366,124	9,107,454	0.26	0.08	773,614	0.02
2017	73	8	16	21,607,605	9,124,174	0.34	0.09	798,511	0.02
2018	68	7	13	22,573,767	9,353,813	0.30	0.07	836,423	0.02

Table 1.5 Air traffic statistics in the Civil Aviation Administration of the Russian Federation during commercial air transportation (based on S analyzes in the Russian Federation)

	Absolute rates						Relative rates					
Year	All commercial transportation			Com. transportation on aircraft with 10 or more seats			All commercial transportation			Com. transportation on aircraft with 10 or more seats		
	ACCID		Death	ACCID		Death	ACCID for 100,000 hours of flight		The dead per 1 million transported.	ACCID for 100,000 hours of flight		The dead per 1 million transported.
	All ACCID	Disasters		All ACCID	Disasters		All ACCID	Disasters		All ACCID	Disasters	
2007	11	7	30	10	6	29	0,5	0,32	0,66	0,46	0,27	0,64
2008	13	8	129	13	8	129	0,55	0,34	2,59	0,55	0,34	2,59
2009	12	6	39	10	4	35	0,58	0,29	0,86	0,48	0,19	0,77
2010	12	5	25	10	4	22	0,51	0,21	0,44	0,42	0,17	0,39
2011	18	10	119	16	8	117	0,69	0,38	1,86	0,61	0,31	1,83
2012	14	7	71	11	6	69	0,51	0,25	0,96	0,40	0,22	0,93
2013	12	5	80	10	4	78	0,42	0,17	0,95	0,34	0,14	0,93
2014	15	7	40	10	3	29	0,5	0,23	0,41	0,33	0,10	0,30
2015	12	5	80	11	4	76	0,42	0,18	0,29	0,39	0,14	0,28
2016	16	6	29	14	4	28	0,59	0,22	0,23	0,51	0,15	0,22
2017	15	7	25	12	5	18	0,51	0,24	0,24	0,41	0,17	0,17
2018	19	9	104	15	6	101	0,6	0,28	0,89	0,47	0,19	0,86

Table 1.6 Comparison of S indicators in commercial aviation of the Russian Federation and the USA for the period 2007–2018

	RF: all commercial transportation; USA: comm. transportation on aircraft with 10 seats or more			Com. transportation on aircraft with 10 or more seats		
	ACCID for 100.000 h	Disasters for 100.000 h	The dead per 1 million transported	ACCID for 100,000 h	Disasters for 100.000 h	The dead per 1 million transported
RF	0532	0.259	0.865	**0.448**	**0.198**	**0.826**
USA	0.324	0.099	0.044	0.324	0.099	0.044
Ratio of rates	1.6	2.6	19.7	**1.4**	**2.0**	**18.8**

The level of S in the commercial CA of the Russian Federation is significantly lower than in the United States. There is a significant lag in relative disaster rates (2–3 times worse) and death rates per 1 million people transported (almost 19 times worse).

Figures 1.1 and 1.2 are graphs of the considered indicators for the Russian Federation and the USA by years.

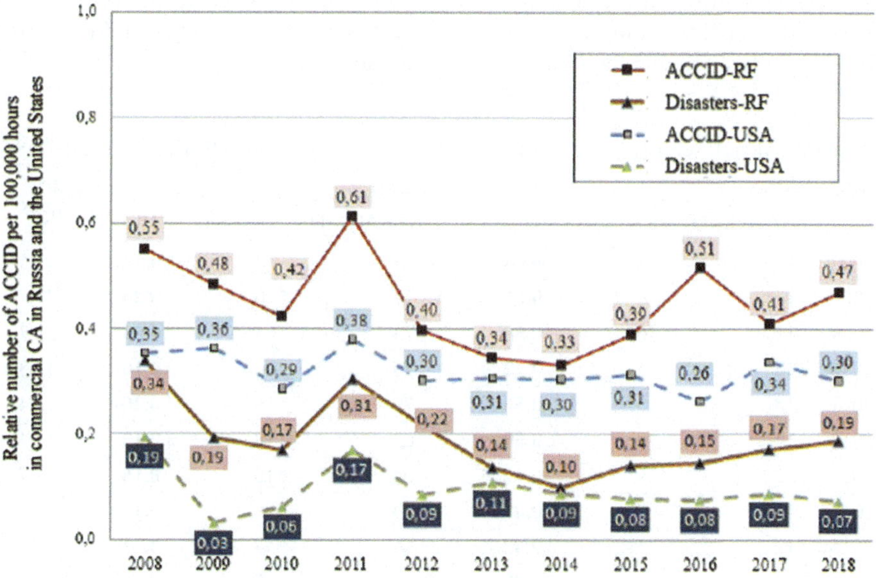

Fig. 1.1 Relative indicators of ACCID for all commercial transport

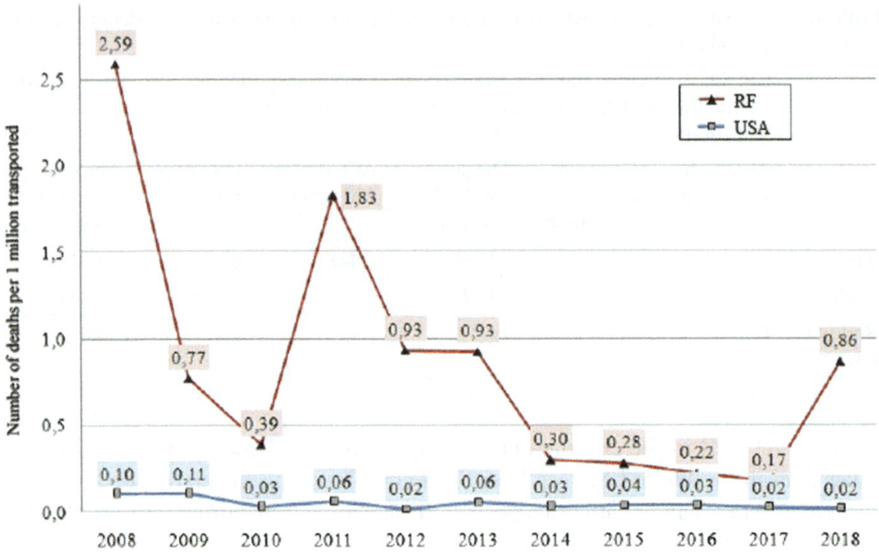

Fig. 1.2 Relative number of deaths on all commercial flights

The calculations of the severity index of ACCID (Dis) according to GOST R-56079-2014 [10] (the ratio of the number of disasters to the number of AE) for the Russian Federation and the USA are given in Table 1.7.

Table 1.7 Aircraft severity index in commercial aviation of the Russian Federation and the USA

Year	RF			USA			D_S^{RF}/D_S^{UAS}
	All ACCID	Disasters	D_S^{RF}	All ACCID	Disasters	D_S^{USA}	
2007	11	7	**0.64**	87	14	**0.16**	3.95
2008	13	8	**0.62**	77	20	**0.26**	2.37
2009	12	6	**0.50**	73	3	**0.04**	12.17
2010	12	5	**0.42**	58	6	**0.10**	4.03
2011	18	10	**0.56**	79	16	**0.20**	2.74
2012	14	7	**0.50**	63	8	**0.13**	3.94
2013	12	5	**0.42**	64	10	**0.16**	2.67
2014	15	7	**0.47**	64	8	**0.13**	3.73
2015	12	5	**0.42**	66	7	**0.11**	3.93
2016	16	6	**0.38**	56	7	**0.13**	3.00
2017	15	7	**0.47**	73	8	**0.11**	4.26
2018	19	9	**0.47 asis>**	68	7	**0.10**	4.60

In Russia, a significantly more significant number of ACCID end in disasters, and more people die in each disaster than in the United States. At the same time, it is necessary to note a high absolute accident rate in general in the USA compared with the Russian Federation. Therefore, in the crash of civil aviation (both commercial and general aviation) in the Russian Federation in 2018, 128 people died, which is perceived as evidence of a low level of air defense. However, in the United States, in 2018, 381 people died in all general aviation crashes.

1.2.2 Comparison of Safety Situation in the Russian Federation and in the Member States of the ICAO

It is recommended that the number of airplanes is assigned per 100 thousand flight hours; however, a comparison with ICAO [11] is traditionally carried out in terms of the number of airplanes per 1 million sorties and only for regular commercial transport. In Fig. 1.3, borrowed from the safety analysis in the CA of the Russian Federation for 2018, the graphs of this indicator for the period from 2008 to 2018 are presented.

We can see that for the period 2013–2017 for regular commercial flights in the Russian Federation, the level of airfare for this indicator was significantly lower than the ICAO average. However, in 2018, there is severe deterioration of this indicator, although when monitoring the indicator for a simple moving average with a smoothing period of 5 years, the indicator remains lower than the ICAO average.

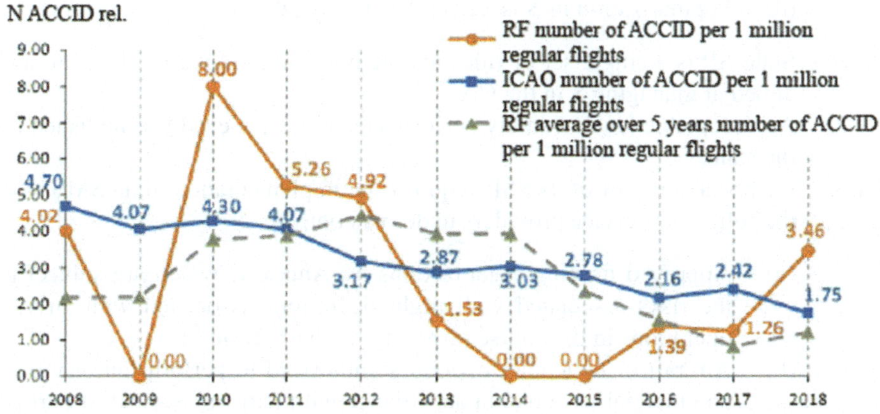

Fig. 1.3 Effectiveness indicators for scheduled flights on aircraft with a maximum takeoff weight of 2250 kg or more (borrowed from the S Analysis in the Civil Aviation Administration of the Russian Federation for 2018)

1.3 Analysis of National and International Regulations Requirements to the Flight Operations Safety Management Systems

1.3.1 ICAO, IATA, and EASA Standards and Recommended Practices

As noted in the International Civil Aviation Organization (ICAO) Global Aviation Safety Plan (GASP) [12], air traffic will double in the next 15 years and, therefore, a proactive approach to addressing existing and emerging S2 risks is needed.

The basis for international regulation of civil aviation, as we know, is the ICAO Standards and Recommended Practices (SARPs) set out in the 1944 Chicago Convention and in 19 Annex to it.

SARPs requirements apply primarily to states; however, for critical operational issues, they also contain requirements for service providers.

Following ICAO Annex 19 [13]:

3.1.1. Each state adopts the State Safety Program (SSP) for the management of safety in the state in order to achieve an acceptable level of effectiveness in ensuring the safety of the civil aviation, which includes the following components:

(a) government policy and objectives for ensuring S;
(b) Risk management for S at the state level;
(c) Provision of S at the state level;
(d) Popularization of S issues at the state level.

State SP is a single set of rules and activities developed by the state and aimed at managing S in the CA.

3.1.2. An acceptable level of safety performance to be achieved is established by the state.

3.1.3. Each state, as part of its SSP, requires the implementation of an SMS with the following service providers under its control:

(a) Approved training organizations by Annex 1, which are subject to the risks associated with flight operations associated with aircraft operations in the course of providing their services;
(b) Aircraft or helicopter operators authorized to perform international commercial air travel in accordance with Part I or Sect. II of Part III of ICAO Annex 6, respectively;
(c) Approved maintenance organizations providing services to aircraft or helicopter operators operating international commercial air travel in accordance with, respectively, Part I or Sect. II of Part III of Annex 6;
(d) Organizations responsible for the type design or manufacture of aircraft in accordance with Annex 8;

(e) Air traffic service providers in Annex 11;
(f) Certified aerodrome operators in accordance with Annex 14.

3.1.4. Each state, as part of its SSP, requires the implementation of an SMS by operators of large-sized or turbojet general aviation aircraft.

3.2 State control over safety.

Each state, following Annex 1, creates and introduces a system that serves as a guarantee that the state will fulfill its safety oversight obligations.

The structure of the SP and SMS of the service provider can be defined as a "conceptual framework" [13].

When comparing the structures of the SP and the SMS of the operator, similarities are revealed. However, the processes regulated by them are fundamentally different, since they are focused on different goals. The goal of the state is to ensure public safety; the goal of the operator is to carry out production work in order to satisfy the customer and make a profit. The state establishes requirements, carries out normative regulation, that is, sets the "rules of the game", and performs supervisory functions.

The main working document for developers was to be the ICAO S Management Manual. In 2013, the third edition of the SMM was published, smaller in volume than previous editions, and more specific. In particular, for the first time, specific examples of calculating the recommended S indicators are given, at several levels, both for the State SP and for the SMS of the operators. At the same time, a careful reading of the document reveals certain contradictions in two parts: Sect. 2.14 and Annex 2. There are differences in translations of the "conceptual framework" in the Russian texts of Annex 19 and the ICAO SMM. Briefly, these shortcomings are analyzed below and discussed in more detail in [14].

The International Air Transport Association (IATA) cites the SMS requirements in the IOSA Audit Guide, September 13, 2019, and the 13th edition is in force. However, these requirements are called ISARPs, although a reservation is made that these Standards and recommended practices apply only to the audit process. Standards, as in ICAO SARPs, are introduced by the "*shall*" verb (shall apply), and recommended practice is introduced by the "*should*" verb.

The introduction notes that, due to the complexity of the task and the presence of specifics of national regulation, the implementation of an SMS can take several years for some operators. Therefore, several ICAO SARPs provisions in ISARPs have so far been recommended practice status.

The requirements for SMS are contained in seven of eight sections of the document (except for SEC—Aviation Safety) related to the main types of operational activities. The production organization (ORG) section contains requirements for the presence of a risk management system, ensuring (confirming) safety, data analysis, and emergency planning, i.e., critical requirements of ICAO SARPs [13]. Several requirements with recommended practice status have become IOSA standards.

A necessary requirement is the availability of a risk management system for all types of activities, as well as SMS training.

The 13th edition took into account the shortcomings of the previous versions, when, except for the mention of risks, the provisions of ISARPs denoted the usual activities to ensure S.

Understanding the needs of airlines for more explicit recommendations, IATA regularly publishes teaching aids and provides training for specialists. We note the 2005 *SMS: The Senior Airline Manager's Implementation Guide*, which aims to combine the recommended practice of providing S with the interests of managers of medium and small airlines. Much attention is paid to the prevention of frequently occurring ACCID, for example, runway incidents [14].

IATA's desire to integrate S, Aviation Safety (AS), and quality issues have been expressed in the Integrated Airline Management System (*IAMS*).

The European Aviation Safety Agency (EASA), as a collective body of the European Community, was established following European Parliament Regulation No. 1592/2002 of July 15, 2002, to establish and maintain a high uniform level of civil aviation safety in Europe.

The European Aircraft Operating Rules (EU-OPS), which replace the JAR-OPS rules, have been amended to comply with the ICAO SMS.

Note that the EASA made an interim decision, which is to develop a common position regarding ICAO with an analysis of the differences. These discrepancies between EASA and the EU states are eliminated during ICAO audits of the Universal Safety Oversight Audit Program (USOAP) during corrective actions.

It is crucial that this document formulates a "cautious" attitude of EU aviation administrations to the designation of an acceptable level in the form of the probability (relative number) of ACCID, and even more so disasters. The reluctance to express the indicator in such risks is explained by the fact that now acceptable levels of S are established only in the UK in CAP–784 [15]. This is a psychological and social issue, and ways to solve the problem require further consideration.

Consider the separate EASA requirements for SMS of aircraft maintenance organizations, as many organizations that perform maintenance of aircraft of foreign production in the Russian Federation have a valid certificate of compliance with EASA requirements (for example, Aeroflot, S7 Technics, Vostok Technical Service, Volga-Dnepr Technik, Rossiya, Gazpromavia, Aurora, a/c Ural, etc.).

The presence of such a certificate is determined by the need to perform maintenance of the aircraft of European registration, as well as contractual obligations with some lessors.

Currently, EASA Part-145 requirements do not oblige aircraft maintenance organizations to implement an SMS. This is apparently due to two reasons.

Firstly, they want to thoroughly work out the requirements and prepare for the long and systematic work on the implementation of SMS in organizations for aircraft maintenance, realizing that this requirement will make significant changes to all activities of the organization.

Secondly, the second edition of Annex 19 and the fourth edition of the ICAO RBM make it necessary to revise the draft requirements considered earlier.

It is worth noting that this "gap" will be filled soon. Confirmation of this can be found in the NPA (Notice of Proposed Amendment) [16], which describes the proposed amendments to the regulatory documents of EASA. This document sets out the requirements for SMS and control of personnel fatigue for aircraft maintenance organizations. In addition to the requirements (Part) themselves, the changes include acceptable Assessment Methodologies Conformity (AMC) and Guidance Materials (GM).

Even though not all service providers have requirements for SMS, EASA has a document (EASA Management System assessment tool v.01–06, September 2017), which contain instructions for evaluating the effectiveness of existing control systems, including SMS. Based on the context, the document may be applicable to any service provider; however, due to the absence of SMS requirements, the aircraft maintenance organizations are not mentioned in this document.

https://www.easa.europa.eu/sites/default/files/dfu/214081_EASA_MANAGEMENT_SYSTEM_ASSESSMENT_TOOL.pdf.

The analysis shows that at present, there are specific differences in the regulatory requirements of the EASA and ICAO Standards, mainly in the direction of tightening the requirements in the EASA documents. Therefore, the questions of the applicability of EASA requirements in the documents of the Russian Federation require consideration in each specific case.

1.3.2 Standards, Regulations, and Guidances of Several States

Long before ICAO, the United States Federal Aviation Administration (FAA) developed a set of regulatory documents and SMS manuals that focused on risk management issues.

Mention should be made of Order 8040.4 circular, Safety Risk Management, June 26, 1998. It establishes the mandatory application of risk management technology by all organizations controlled by the FAA. The risk management process is required to be clearly defined and documented. Responsibility areas for the process should be established.

Based on the circular, specific guidelines were developed, for example, the FAA System Safety Handbook (2000), the System Safety Management Program for ATS and ANO organizations, introduced back in 2001, and the SMS Program 2004 [16].

In July 2006, the FAA published Advisory Circular AC-120-92, which is, in fact, the SMS standard for all US operators. As evidence of this status of the document, we cite the following provision: "The SMS standard contained in Annex 1 to this Advisory Circular (AC) includes the requirements for S. The content of the document is similar to international standards for quality management, environmental protection, operational safety, and health protection."

As described in [17], the document was developed as a research project under a contract with the FAA Technical Center. The work began after an analysis of the requirements by the group's specialists with the participation of representatives from the FAA, major airlines, and the Air Line Pilots Association (ALPA). A study was carried out of all the developments on existing S support systems in countries that have successfully implemented SMS: Canada, Australia, New Zealand, and Great Britain. The results of the work of modern theorists in the field of S, first of all, Reason [18], are summarized, and the military standard *Mil-Std* in force in the USA is taken into account.

The format of the International Organization for Standardization (ISO) was taken as the sample, mainly the ISO 14001 environmental standard. In developing the requirements for safety processes, the ISO 9000 standard was also used. In the circular for the first time, a functional diagram of the SMS construction was given, which ICAO as a "conceptual framework" for both the government program then adopted and the SMS of the service provider (see Sect. 1.3.1).

The Standard must be functionally oriented; that is, its requirements describe the expected processes, and not how they will be implemented. This provides operators with the broadest possible opportunity to build programs so that they are combined with existing or future business and management models. The document emphasizes the critical role of S risk management in SMS and describes the requirements and process clearly and concisely. The process is consistent with ICAO recommendations, but the risk matrix is different from the known matrices. The advantage of the FAA is that it provides definitions of key terms.

Thus, it must be recognized that the FAA did not limit itself to recognizing SARPs, using the ICAO SMM, and encouraging operators to use it in their activities, but developed its document, which is a state standard and a guide for building SMS in the industry and at the level of operators.

FAA's current SMS policy is consistent with ICAO SARPs, as documented:

1. FAA Order 8000.369B, Safety Management System
   https://www.faa.gov/documentLibrary/media/Order/FAA_Order_8000.369B.pdf
2. FAA Order VS 8000.367, Aviation Safety (AVS) Safety Management System Requirements 05/09/17.
   https://www.faa.gov/documentLibrary/media/Order/VS_8000_367B.pdf
3. AC No: 120-92B Date: 1/8/15a
   https://www.faa.gov/documentLibrary/media/Advisory_Circular/AC_120-92B.pdf.

However, not all service providers are required to do so. For example, FAA air law does not establish requirements for the availability of SMS for aircraft maintenance organizations, but there are recommendations for the implementation of SMS voluntarily.

The same recommendations apply to those service providers from whom the FAA requires an SMS (for example, for operators).

The mentioned recommendations cover in detail the development, implementation, and continuous improvement of the SMS of the service provider. SMS recommendations are also provided.

The annex of FAA's best practices in terms of regulatory regulation of S management issues in the CA of the Russian Federation may be limited. In addition to the lack mentioned above of requirements for the organization's SMS for maintenance, there are other differences, for example, differences in the classification of aircraft, different approaches to the use of external control as a source of information about hazard factors, especially regarding the use of tape recordings in the cockpit.

Back in 2001, before ICAO and other countries, the United Kingdom **Civil Aviation Authority (CAA)** issued a Manual on the Safety Management System for Airlines, containing all the most critical provisions of SMS, including a description of the risk management system [19].

In 2002, an SMS manual for air traffic management SAR-730 "*Safety Management System for Air Traffic Management. A Guide to Implementation*" was published. This document also guided the implementation of a risk management system. The peculiarity of these two new documents regarding risk management was the use of a risk matrix of 4 × 4 size.

In 2006, in addition to CAP 730, CAP-760 [20] was published, the title of which translates as "Guidance on the identification of hazards, risk assessment, and compilation of a list of hazards for organizations operating aerodromes, aerodrome services, and air traffic services providers". As the name implies, this document is specifically designed to standardize the risk management process.

Since 2008 SAR-712 was canceled, a similar standard can be expected for airlines. So far, instead of SAR-712, the document "Guidelines for Organizations" [139] is used. In this document, as in SAR-760, the 5 × 5 matrix already adopted by ICAO is used, but the probability assessment uses its approach and original assessment criteria.

A feature of the document [21] is the inclusion in it of the interaction scheme of the SMS and the quality management system.

A brief review of the documents shows that the UK aviation authorities are systematically developing, implementing, and updating SMS requirements for their operators, as well as providing sufficiently detailed guidelines for organizing and executing processes, including safety risk management.

The introduction of SMS in Canada began in 2002 when the Department of Transportation of Canada announced its desire to reduce its oversight role [22]. The decision of the Ministry of Transport of Canada on the mandatory implementation of SMS was adopted in 2005. SMS is presented as a system for identifying, describing, combining, managing, eliminating, and tracking risks; a "roof or umbrella" covering existing programs.

Requirements for the presence of an SMS for all participants in the air transportation market are included in the Canadian Aviation Rules (CAR):

CAR.106, CAR.107.3—for operators (airlines);
CAR.302—for aerodromes and airports;

CAR.406.76, CAR.426.76—training and licensing;
CAR.573.16, STD.573.16—for maintenance organizations;
CAR.720.35–720.38—for ground handling;
CAR 805–CAR.825—ATS and air navigation support for flights.

The goal of SMS implementation is to strengthen industry responsibility, introduce a consistent S culture, and help improve operators' performance. SMS in Canada is also called the "first significant effort to standardize S programs" or the "move toward self-regulation".

Following CAR.106, the following is required to be included in the airline FOM (Flight Operations Manual):

- a risk management plan containing the procedure for appointing the responsible person, the roles and responsibilities of personnel according to the quality program, the goals, and means of assessment, the policy of internal reports on dangers, incidents, and aircraft Accidents (ACCID), including the conditions of impunity, and SMS audits to evaluate effectiveness;
- procedures for reporting on hazards, incidents, and emergency procedures;
- procedures for the collection of data on hazards, incidents, and emergency procedures;
- audit system;
- the rules of data analysis during the audit and the adoption of corrective measures;
- requirements for the training of the flight manager, maintenance manager, and personnel on their responsibilities to SMS;
- procedures for regular and urgent reports to responsible executors.

In 2004, the Canadian Department of Transportation issued the Risk Management Guide [23], and in May 2005, a detailed Guide to assessing SMS in airlines. This Guide provides specific provisions for evaluating the operation of SMS, and defines the audit procedure, documents, and procedures, the presence of which is checked during the audit. The list of questions that the auditor will ask top managers, managers, and ordinary employees of the airline to make sure that the system works is given. These questions cover all parts of the SMS and are selected in such a way that they give a good idea of the functioning of the system and employee awareness. The rating system is flexible and allows us to set several levels of SMS development and suggest ways to improve it.

Canada is a country with one of the highest levels of S. The aviation authorities have developed and implemented SMS as a standard. A significant number of guidelines for the development and implementation of SMS for operators have been prepared and widely used.

As in the UK, Canada has developed detailed instructions for the implementation of SMS in airlines of various service providers, which can be used in the development of relevant documents of the Civil Aviation Administration of the Russian Federation.

1.4 Description of the Current Situation in the National Safety Program and in the Flight Operations Safety Management System of the Russian Federation

As shown in Sect. 1.1, in the regulatory documents and standards of the Russian Federation, there is no complete clarity with the definitions and approach to solving safety management problems. This situation also affected the situation with the regulatory regulation of this issue in civil aviation.

Following Article 2 of the Air Code of the Russian Federation (AC RF), the Air Law of the Russian Federation consists of the Air Code, Federal laws, decrees of the President of the Russian Federation, decrees of the Government of the Russian Federation, federal rules for the use of airspace, and federal aviation rules, and also adopted following them other regulatory legal acts of the Russian Federation.

In accordance with the new Article 24.1, three requirements are introduced in the RF AC:

1. The implementation of the state system of safety management is ensured in the Russian Federation following ICAO Standards.
2. Service providers are required to have an SMS according to ICAO standards.
3. The authorized body in the field of civil aviation collects and analyzes data on hazard and risk factors that pose a threat to the safety of flights of civil aircraft, stores these data, and exchanges them by ICAO standards.

The practical implementation of these requirements is problematic. Let us dwell briefly on the current situation.

1. *The implementation of the state system of safety management is ensured in the russian federation following icao standards*

Since it speaks of ICAO Standards, it is necessary to use the ICAO terminology [13] and talk about the State Aviation Safety Program (SSP).

In 2001, "The Fundamentals of the Policy of the Russian Federation in the Field of Aviation Activities for the Period Until 2010" were adopted (Approved by Decree of the President of the Russian Federation 03.03.01, No. 241). In 2002, The Federal Target Program "Modernization of the Transport System of Russia (2002–2010)" was adopted (approved by Decree of the Government of the Russian Federation of 05.12.01 No. 848). During this period, approaches to the management of air safety from the perspective of risk management have already been developed in the global aviation community. On the other hand, several laws and standards were adopted in the Russian Federation, in which the problems of technological safety were prescribed to be solved based on risk management. However, neither in the Policy nor in the Federal Target Program, there is a mention of the risks of S.

The State power supply support program [24] occupies a special place in the power supply system. The Program aims to eliminate problems and implement SMS in civil aviation. Consider how the program is supposed to solve this problem.

The program determines the content of the problem and indicates the need for its solution by software methods. It is stated that S management requires planning organizational measures for identifying and eliminating the risks of aircraft accidents, interaction on the prevention of airspace for all participants of the aviation transport system during operations, their provision, and investigation of ACCID. It is stated that an acceptable level of S should be established, which should become the basis on which S will be evaluated by the authorized body.

To achieve an acceptable level, several tasks are being solved, among which is the creation of a power supply management system, as well as providing a systematic approach to identify sources of danger and control risk factors to minimize the number of human casualties, as well as the size of material, environmental, and social damage.

The program determines the sequence of solving the set of tasks. At the first stage, it was envisaged "the development by federal executive bodies of federal and departmental target programs (lists and plans within the budget allocations provided for the fulfillment of obligations), as well as requirements for the implementation of international standards and procedures for interagency consideration of problems of ensuring S for decision-making and events to prevent ACCID". This situation is not very clear and not specific. There is no information on the situation with the development of programs in available sources of information.

In the annex to the Program in Sect. 1.2 was planned to create "rules of development and scheduled to ensure an acceptable level of S", in paragraph 5 of Annex "establishing acceptable levels of S and S indices in the operation of aircraft, subject to air traffic services and aerodromes". Currently, these plans are not implemented.

2. *The obligatory presence of SMS in air operators*

Known Policy the Federal air transport Agency dated 25.11.2009 №. GK1.22-2979 on the mandatory implementation of SMS in airlines required to develop and implement SMS and to prepare the relevant SMM in a month. Formally, it was done, and all the major airlines have made their SMM of the General Provisions of the ICAO SMM, but experts know that most of these manuals were not working documents.

Through the work of the Interdepartmental Commission on the S annual monitoring of the implementation of ICAO standards, aircraft operators (airlines) are proposed to be done in accordance with the procedures of the guidelines, ICAO Doc. 9734 AN/959 and Doc. 9735 AN/960. However, Doc. 9735 "Guidance on conducting audits of oversight safety" is a guide for the experts of ICAO, state-controlling bodies. Doc. 9734 "Guidance on monitoring of the safety of flights" contains guidance material on the establishment and functioning of the state system of assuring S and not of much help in the formation of requirements of the operator's SMS. Direct requirements for the operator are contained only in paragraph 2.3.5 of this document, "Responsibility of operators and service providers for safety".

3. *Collection, storage, and exchange of data on "hazards and risk factors"*

This requirement is not entirely clear, because it is not defined what is understood here as a "hazard factor" and a "risk factor". In the Russian text of ICAO Annex 6, which was in force at the time of the preparation of the amendments to the VK, the term "hazard factor" corresponded to the term "hazard" in the English text and was defined as "the condition, facility, or activity potentially causing bodily harm to personnel, damage to equipment or structures, material losses, or reduced ability to perform the prescribed function".

As for the term "risk factor", its appearance in ICAO documents in Russian is explained by an error in the translation of the ICAO SMM, second edition, where the term "risk factor" translates the English "*safety risk*".

In order to understand the "collection, storage, and exchange" of what data is discussed in the AC, you need to understand the meaning of these terms clearly.

The same issues of terminology that affect the understanding arise in the study of **the Resolution of the RF Government dated 18.11.2014, No. 1215** "On the procedure of development and annex of safety management systems of aircraft, and collect and analyze data on hazard and risk factors that threaten the safety of civil aviation operations, storage and sharing". Therefore, in the Resolution given differently from ICAO SARPs definitions of SMS and the danger factor instead of the "maintenance provider" uses the term "service provider".

Note also, that at the time of entry of Resolution into force, the real steps for its implementations were not determined by RF Ministry of transport.

As we know, many States maintain a database of aviation events (in Russia, it is ASEBP) and have an established procedure for the provision of event data (for example, ADREPICAO), and there is data exchange system between operators (for example, STEADESIATA). This experience can be useful in implementing said Resolution.

In the FAR-128, the operator introduces an SMS which

- includes the processes of determining the physical and potential threats to safety and assesses the risks involved;
- provides corrective actions necessary for keeping the safety indicators;
- provides constant monitoring and regular assessment of safety performance;
- provides continuous improvement of S.

Here are mentioned the concept of risk, but not explained.

If we approach FAR-128 and others FAR to Technical regulations (which conform to the requirements of the Federal law "On technical regulation"), it is appropriate to give here the principles desirable to build a common technical regulation [25]. It is believed that should be based on the requirements of focus and consistency. This means that the structure of each such document should have the following components:

- starting position (concept, i.e., the essence of the regulations);
- methodology (the skeleton or framework requirements);
- technology (model) of safety in a particular area, as well as the preamble and final clauses.

In [25], we determined the semantic load and the approximate contents of the above standard technical regulations.

Technical regulations for S, corresponding to these qualities, are there in Russia. There is no state standard for safety.

Currently, the GOST series is being developed under the general name "Air Transport. Aviation Safety Management System". These GOSTs introduce new concepts: "Safety of Aviation Activity (SAA)", "management of SAA", "risks of aviation activity", etc. It is assumed that the system of these GOSTs will regulate safety management at the level of the entire industry.

Safety of aviation activities is considered as the general safety of the industry, including the following aspects: safety, aviation transport safety, industrial safety, environmental safety, and information safety.

This approach is justified by the desire to provide service providers with requirements, as well as methodological material for the development of integrated safety management systems of organizations, aviation service providers, which corresponds to the modern approach. However, it must be remembered that the explicit requirements of ICAO SARPs and regulatory documents of the Russian Federation on the development of safety management systems apply today only to the development and implementation of SMS. Accordingly, the remaining requirements of the developed GOST will be recommendatory.

The development of such GOSTs will require additional research. For example, the industry has repeatedly raised the question that the concept of "safety" as the Russian analog of the concept of "*safety*" is limited. Some airlines [26] use the concept of "operational safety" based on "*operational safety*" from IATA documents. There were even suggestions in general to replace the term S with another.

However, the concept of "safety" in the broadest sense [27–33] has become so firmly entrenched in regulatory documents and in Russian texts of ICAO Annexes that, in our opinion, the terminology should not be changed. In support of this, we give the S definition from Annex 19 as a condition "in which the risks associated with aviation activities related to the operation of aircraft or directly ensuring such operation are reduced to an acceptable level and are controlled".

An outstanding aircraft designer, the academician of the RAS G.V. Novozhilov, expressed in his book [34]: "safety is a property of the entire aviation transport system."

Indeed, if we focus on the requirements of Sect. 3.1.3 of ICAO Annex 19, then all service providers for which an SMS is required are included in the circle of organizations involved in "aviation activities".

From the above brief analysis, it is clear that when developing new GOSTs, issues of terminology should be given increased attention.

In general, it is necessary to recognize that, as of the present, the regulatory documents of the Russian Federation on S and the State S program need serious revision and updating. In particular, there are no approved definitions of the concepts: "safety", "safety risk", "acceptable risk", "risk management", and "danger factor (threat, danger)", and the requirements for the operator's SMS are not specific enough.

Inaccuracies impede the implementation of SMS in the translation of ICAO S documents into Russian. Consider the inaccuracies of the transfer to the ICAO SMM 2013. Compare safety definitions.

Safety. *The state in which risks associated with aviation activities are reduced and controlled to an acceptable level.*

Безопасность полетов. *Состояние, при котором риски, связанные с авиационной деятельностью, снижены до приемлемого уровня и контролируются.*

The translation is not precise. In Russian, "to control" means to exercise control [14]. Synonyms: to oversee, verify, inspect, observe, make sure, and compare. To control does not mean to control (to control). The English definition states that risks are managed so that they are no higher than an acceptable level, which is the essence of SMS. It should be noted that such a definition was in the ICAO SMM, 2009. What explains the change is not clear because the ICAO concept has not changed.

Definition of S indicators.

Safety Performance. *A State or a service provider's safety achievement as defined by its safety performance targets and safety performance indicators.*

Эффективность обеспечения безопасности полетов. *Достигнутый государством или поставщиком обслуживания уровень безопасности полетов, определяемый установленными ими целевыми уровнями эффективности обеспечения безопасности полетов и показателями эффективности обеспечения безопасности полетов.*

In Russian translation, the phrase "**safety effectiveness**" appeared which changed the meaning of the definition.

Proposed Translation Safety Level: The level of S achieved by the state or service provider, and determined by their target S levels and S indicators.

The proposed transfer Level of safety: Made by the state or provider level S, and defined by their established target levels of S and S indicators.

Corresponding changes are also proposed in the translation of the terms: "Safety performance indicator" and "Safety performance target".

Compare the definition of risk.

Safety Risk is the projected likelihood and severity of the consequences or outcomes of a hazard.

Risk factors for safety are the predicted probability and the seriousness of the consequences or results of the implementation of threat factors.

Instead, the correct translation of the term as a "safety risk" was coined "risk factors". Hereafter SMM 2013 in paragraph 2.14.2 probability and seriousness together is a "risk factor" in the singular, and then stated "likely risk factor", i.e., "probability of probability"!

Differently, translate the English "hazard": in Annex 6 long years it was the "danger factor", in Annex 19 [13] is a "source of danger", although published in the same 2013 ICAO SMM [35] is a "threat factor".

The English wording of the four-component and 12 elements of SMS are the same and in Annexes 6 and 19, and all three English editions of the SMM 2006, 2009, 2013, 2018, Russian texts of these "frames" differ in all of these documents. Therefore, the translation in the SMM fourth edition 2018 has six differences from the translation of the same text in Annex 19 (see Table 1.8).

In Annex 19 excluded "risk factors" in category 2 but in Sect. 3.1 added "the efficiency of the S", which is not in the English text, and was no sooner in Annex 6 (see above).

Table 1.8 Differences in the wording of the SMS framework

1. The policy and objectives of safety: 1.1 Management commitment. 1.2 Accountabilities and responsibilities in the field of aviation safety. 1.3 Appointment of key employees responsible for safety. 1.4 Coordination of planning measures in the event of an emergency situation. 1.5 SMS documentation. 2. Risk management safety: 2.1 Identification of sources of danger. 2.2 Assessment and risk reduction safety. 3. Flight safety: 3.1 Monitoring and quantitative assessment of the effectiveness of safety. 3.2 Implementation of the changes. 3.3 Continuous improvement of SMS. 4. Promotion of light safety issues: 4.1 Training and education. 4.2 Exchange of information on safety.	1. Policy and objectives in the field of aviation safety: 1.1 Commitments and responsibilities of leadership. 1.2 Responsibilities of safety. 1.3 Appointment of key personnel safety. 1.4 Coordination of planning efforts in the event of an emergency. 1.5 SMS documentation. 2. Risk management safety: 2.1 Identification of hazards. 2.2 Assessment and reduction of risk factors for safety. 3. Flight safety: 3.1 Monitoring and performance measurement of aviation safety. 3.2 Management of change 3.3 Continuous improvement of SMS. 4. Promoting flight safety: 4.1 Training and education. 4.2 Dissemination of information on safety.

1.5 Safety Risk and Its Acceptable Level

1.5.1 Analysis of Different Approaches to Risk Definition

The concept of risk is used in a wide range of Sciences [3, 22, 36–38]. Research on risk analysis can be found in the literature on psychology, medicine, and philosophy. In [39], it is argued that "risk" in modern science and culture is a broad and generalized category like "meaning", "value", "justice", etc.

In the theory of reliability, the risk is the probability of failure [40–42]. In the context of deviation from the target risk is the possibility of a negative deviation of the results of operations from expected: "the effect of uncertainty on objectives" [43], "undesirable option" [44]. Catastrophe theory uses the term "risk" to describe accidents and natural disasters [45].

In practice, there are a few basic concepts of risk [1, 38, 46–49]:

- the risk characteristics of potential danger as the possibility of realization of adverse events that cause harm;
- risk as a measure of discrepancy between the results of the decisions from the point of view of their usefulness, and as an opportunity to implement inferior alternatives;
- the risk as to the relationship between loss and profitability, including the concept of "chance" gets an unplanned benefit.

Risk management, the main S, is usually the first of these concepts that correspond to the ICAO approach but in the real production activity of the airline used and the other two.

The concept of risk management is given in Chap. 2, Fig. 2.1.

The most significant methodological problems are associated with the identification of hazards and risk assessment. Since S is determined by the condition in which the risk of S is maintained at an acceptable level, we must be able to assess the risk value and compare it with the specified acceptable level.

Although according to Kelvin (W. Thomson), "everything that exists, exists in some quantity and can be measured" [36, p. 33], in the case of risk, measurement is difficult due to the direct connection of the concepts of risk and uncertainty. Risk is essentially a measure of the uncertainty of the outcome of an activity, and in this case, a measure of the unwanted completion of a flight. Uncertainty in the manifestation of Hazard Factors (HFs) that affect the outcome of the flight, and, accordingly, determine the risk, is different.

In Fig. 1.4, built based on [38], the main classes of HF are shown, and Table 1.9 shows examples of HF in the form of flight environment conditions.

The work of Professor Zhivetin [50] provides a peculiar understanding of risk and safety in the activities of civil aviation. S is interpreted as an integral characteristic of three parameters: risk, cost of the event, and degree of acceptability, and risk means the probability of an event. The "risk classification scheme" presented by him [50, p. 22] is not consistent with the ICAO approach.

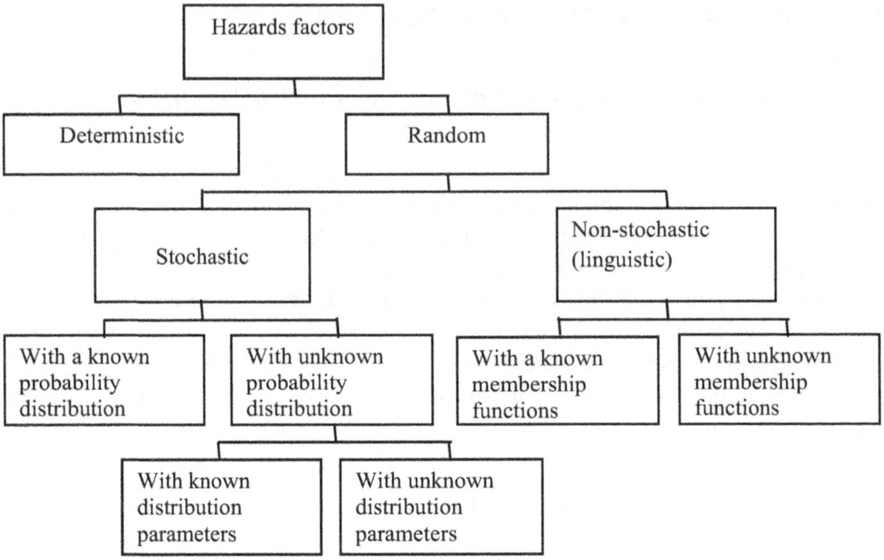

Fig. 1.4 Classification of factors affecting flight outcome

Table 1.9 Examples of environmental factors of varying degrees of uncertainty

Certainty conditions (deterministic)	Conditions for partial uncertainty (stochastic)		Conditions for complete uncertainty (linguistic)
	Probabilistic	Statistics	
Complicated terrain and obstacles	Failures in ATC systems	Collisions with birds	Disadvantages of the rules of documentations
Limited airfield sizes	Failures of ground and space hardware	Extreme weather conditions	Disadvantages of training and training programs
Lack of airdrome radio and lighting equipment	Distortion of aeronautical data	Weather hazards	Personnel errors during maintenance, flight organization, ATC, etc.
Flight in special geographical conditions	Entering the satellite track of another aircraft	Errors in weather forecasts	Failures in aviation safety
Constant restrictions		Dangerous geophysical phenomena	
Lighting conditions			

Analysis of various definitions of safety risk shows that the use of two groups of key terms is common:

- "probability", "opportunity", "possible danger", "threat", "risk factor", "danger factor" characterizing the randomness of the impact of a negative factor, the probability, degree of possibility, or frequency of occurrence of an adverse event;
- "losses", "damage", "deviation of the result from the planned", related to the consequences of this event.

Thus, it is recognized that a safety risk is a category defined by two components: the probability (possibility) of a hazardous event and the consequences of that event.

The most developed and used for risk assessment of hazardous facilities is the so-called "**technocratic concept**", enshrined in many GOSTs. For example, the following brief definitions are given in GOST "Safety Aspects" [51]:

Safety: There is no unacceptable risk.
Risk: The combination of the likelihood of damage and the severity of the damage.
Damage: Physical damage or other harm to human health, or damage to property or the environment.
Danger: Potential source of damage.

These definitions in meaning coincide with the definitions of the ICAO SMM [35], but suggest a quantitative assessment.

The product of the probability of each *j*th event and the damage can be considered as the risk associated with this event:

$$R_j = P_j \cdot S_j \tag{1.1}$$

Given reasonable estimates of the probability distribution of damage, a probabilistic risk assessment model can be built.

It is this approach to the general risk assessment that is proposed in the Article [44] for any type of risk—from planetary to personal. At the same time, the use of nonparametric distributions is recommended, which are more relevant to the real world [52].

This approach is fully justified only for assessing risks associated with individual factors and stochastic nature.

1.5.2 Application of Fuzzy Sets and "Soft Estimation"

As for factors of a non-stochastic nature, purely probabilistic methods are not always applicable [6, 36, 37, 49, 53–57]. In this area, especially for rare events, with "heavy tails" of distributions and a probability of "almost zero", other methods are used based on the Theory of Fuzzy Sets (TFS), "soft estimation", interval mathematics, etc.

A slightly different interpretation of risk is used. For example, Professor Kuklev [58, 59] offers an approach that allows a dual interpretation of the concept of "risk" both as a measure and as an event.

Two possible risk measures are identified:

- a measure of the first kind, evaluating the possibility of a risky event;
- a measure of the second kind denoting an immediate, measurable danger.

An event is considered risky if the necessary safety cannot be provided with the available technical means. A risk is an event in the form of a chain $A_i = L(qi_o, qi_2, \ldots, qi_3)$, where qi_j—chain elements. Ultimately, the overall risk of danger is the sum of the risks of the elements:

$$\mu_{R2} \sum \beta = \sum_{i(\beta)} m_i(P),$$

where β—event number;

P—probability distribution on which particular risks depend on m_i.

It is known that using TFS, it is possible to formalize and study many quantitative and qualitative concepts and estimates by presenting them in the form:

$$\forall x \in X \quad A = \{(x, \mu_A(x))\},$$

where $(x, \mu(x))$—a pair of components made up of elements x of set X and its membership function;

$\mu_A(x)$—membership function or degree of membership of x to set X, expressed by a real number in the range $M = [0, 1]$.

Membership function $\mu_A(x)$ is considered as the distribution of the possibilities of the appearance of specific real numbers. The modal value x_m of a fuzzy quantity x is an element of the subset with a unit degree of membership $\mu_A(x_m) = 1$.

A fuzzy number is considered to be an upper semicontinuous fuzzy interval with a convex membership function and a single modal value. This concept is expressed in practice by the words "approximately, roughly, around x_n".

Using TFS, models are constructed for assessing the risk of rare and most severe disasters and accidents in systems with uncertainty such as "human–machine environment". For the preparation of fuzzy estimates and "soft assessment", particular technologies and programs have been developed [60–62], and qualified experts are involved [52, 63–65].

The use of the results of a new direction—cognitive technologies [55]—looks promising.

However, in our opinion, it would be wrong to contrast the probabilistic approach and the TFS approach for practical use in managing safety risk for the following reasons.

1. Probabilistic models are well developed in the theory of reliability [40, 42], and have long been successfully applied in various fields, in theoretical and practical developments. The entire system for assessing and maintaining the airworthiness of aircraft, evaluating the suitability of functional aircraft systems based on the results of the analysis of the types and consequences of failures, and replacing removable elements is based on probabilistic reliability assessments [27, 28, 30, 31, 66–69].

Probabilistic methods are the main ones in the Aviation Risk Management Studies of the Netherlands Transport Safety Institute (NLR) [70, 71].

Many of the risk assessment methods used in different fields, enshrined in GOST RF, are based on probabilistic estimates.

2. As shown in the works of Professor Orlov [72, 73], TFS can be reduced to the theory of random sets and, thereby, to probability theory. Interval estimates, as shown in [44], can also be reduced to probabilistic if we assume that within the fixed intervals, the random variable is distributed evenly.
3. Probability theory and probabilistic estimates (in the sense of the axiomatic definition of Kolmogorov [74]) are successfully used to study the risks of major technological accidents and natural disasters with power statistics $p(x) \approx x^{-a}, a \geq 1$, those being "heavy-tailed distributions".

There are known works on probabilistic estimates of events with extreme distributions in the analysis of the safety of critical objects of a correspondent RAS Makhutova [75–77].

Probabilistic Safety Analysis (PSA) is successfully used at nuclear facilities and in the oil and gas industry [36, 49, 78, 79]. The well-known concept of six sigma [80] also works on probabilities close to zero.

4. Expert assessment, as one of the fundamentals of the use of TFSs, is not fundamentally new in the theory of safety. Methods of expert assessment for objects and processes of non-numerical and non-stochastic nature are well developed and have long been used [64, 81, 82].

On the other hand, examples of the specific practical application of TFS in the field of S inaccessible sources could not be found. Very conditionally, this direction includes the program for assessing the risk of collision of a working aircraft with the ground CFIT [83], the FORAS program [84], and the "risk matrix".

5. The modern approach to safety theory involves the synthesis of various methods and tools. In the fundamental series "Safety of Russia" [1, 8], TFS-based methods occupy a significant place and are considered, along with other methods, as an integral part of the general risk management methodology. For example, when using the classical PSA approach in event/fault trees, individual probabilities, especially by personnel reliability, can be calculated using TFS.

The book explores the applicability in SMS of various risk assessment methods. Some of them were used directly in the development of new methods in the framework of the proposed methodology.

In particular, TFS and "soft assessment" are used in the following provisions:

- When developing an algorithm for Bayesian refinement of the probabilities of events in a system based on trees of aviation events;
- In the formation of a fuzzy logical conclusion for calculating the "indicator of the degree of possibility" in the development of a risk management method using conditional indicators;
- When calculating the "Risk Ratio for Defections and Events" (RRDE);
- When assessing the significance of indicators in the method of factor analysis of safety performance indicators.

1.5.3 Concept of Acceptable Risk

According to modern concepts, there cannot be absolute safety, and there is always a residual risk.

In GOST [85, p. 3.7], the acceptable risk is defined as "the risk that in this situation is considered acceptable given existing public values". And further: "The acceptable risk is the optimal balance between safety and the requirements that a product, process, or service must satisfy as well as factors such as profitability for the user, cost-effectiveness, customs, etc."

NASA Guidelines [57, p. 10] defines acceptable risk as "a risk that is understood and agreed upon by the project/program management body and the customer as acceptable to achieve a certain success criterion within the approved amount of resources".

Allowable risk is achieved through an iterative process for assessing and mitigating risk. Acceptable, maximum permissible, regulated risk is risk measures characterized by goals of achievement and ways of controlled change in the level of risk. Moreover, as noted in [86, p. 74], "the acceptable value of risk is derived from the level of the country's economy. The higher the level of the economy, the lower the value of the acceptable (permissible) risk".

In Fig. 1.5 is a breakdown of the "scale" of the risks of the impact of economic activity on the population based on the experience of such countries as the Netherlands, the USA, and Great Britain.

It is noted that for the development of legislation in this direction, it is required to establish a more detailed and complete classification of emergencies, to determine the criteria for their relationship to a particular class.

It is indicated that the central element of the safety policy is risk management, and "the regulatory organization must supervise and control the level of risk until, in the process of reducing it, it reaches a negligible level according to the accepted three-zone 'scale' of risks" [8].

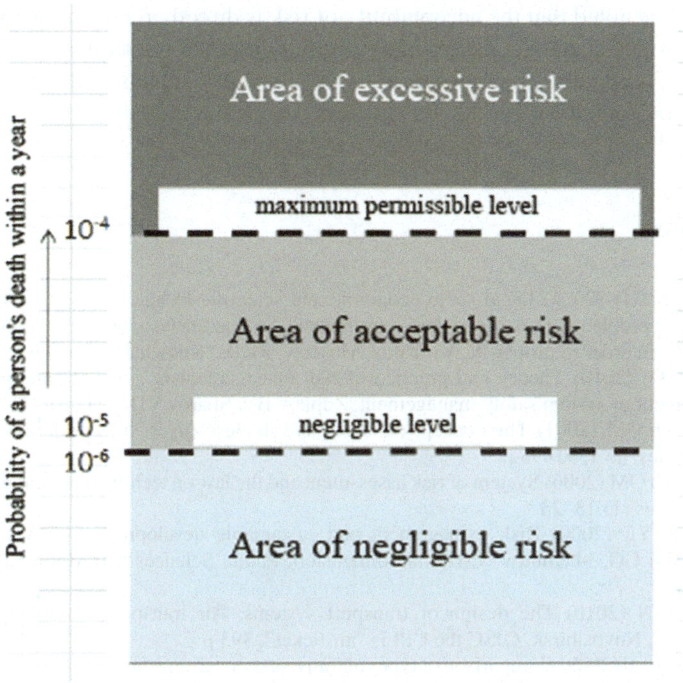

Fig. 1.5 Classification of areas of economic activity by the value of the individual risk of death (based on [8])

This approach is consistent with the principles of ICAO safety risk management at the state level.

In the Russian Federation, at present, there are no legislatively established criteria for acceptable risk; however, several specialized scientific organizations have carried out studies on the analysis of methods for assessing and managing natural and technological risks for more than 10 years. In a generalized form, these results are given in [8], and it can be assumed that risk levels will be established on their basis.

For example, for nuclear fuel cycle enterprises, risk values are recommended as the probability of death per person per year:

- Company personnel—1×10^{-5};
- Population in the sanitary protection zone—1×10^{-6};
- The rest of the region's population—1×10^{-7};
- The population outside the region—1×10^{-8}.

These recommendations can also be used to establish acceptable risk levels for transport, in particular for CA.

It should be noted that the acceptability of risk is directly related to its perception in society. So the society considers single, but with grave consequences, events (air crashes) less acceptable than a large number of small incidents with the same level of risk (traffic collision).

References

1. Russia's Safety (2006) Legal socio-economic and scientific-technical aspects. Risk analysis and safety problems. In 4 parts. Part 1 framework for the analysis and regulation of safety. Scientific. in order to proper. K. V. Frolov. Moscow: MGF "Knowledge", 640 p
2. Zubkov BV (2010) Theory and practice of risk management in aviation enterprises in the development of system safety management. Zubkov BV, Sharov VD – M.: MGTU CA, 196 p
3. Kovalevich OM (2001) The concept of "risk" and its derivatives. In: Problems of safety in emergencies, no 1, 91–98 p
4. Kovalevich OM (2006) System of risk assessment and the law on technical regulation. Saf Prob Emergencies (1):13–23
5. Vorobyov YL (2000) Risk management and sustainable development. In: Vorobyov YL, Malinetskii GG, Makhutov N. Human dimension. Public Sciences and Modernity, no 6, pp 150–163
6. Plotnikov N (2010) The design of transport systems. Air transport: monograph/N. And. Carpenters. Novosibirsk, CJSC the CPI is "air ticket", 393 p
7. Barosh JS et al (2009) Comparison of types of transport concerning the sustainable development of society. Bulletin of the Dnepropetrovsk National University of railway transport named after academician V. Lazaryan, 210–214 p
8. Russia's Safety (2008) Legal socio-economic and scientific-technical aspects. In: Makhutov NA, Pulikovsky KB, Shoigu SK (eds) Risk analysis and safety management. (Methodical recommendations). MGF "Knowledge", Moscow, p 672
9. NTSB USA [Electronic resource] access mode. http://www.ntsb.gov/data/aviation_stats.html
10. 23743-88 GOST B, GOST In 20570–88, GOST In 20436-88 Safety, reliability, testability and repair processability (1988) Goskomstat SSSR, p 10
11. Global Aviation Safety Plan (2007) ICAO, 25 p
12. Global Aviation Safety Plan (2013) ICAO, 76 p
13. Annex 19 to the convention on international civil aviation. Safety Manag. ICAO, 44 p
14. Sharov VD (2013) About some problems of understanding the terms related to safety management. In: Balls VD. Safety problems, no 9, S.12–S.21
15. CAP 784 State Safety Program for the United Kingdom. CAA UK 2009
16. System Safety Management Program, FAA (2004)
17. Arend D (2006) Concept of safety management system embraced by many countries. ICAO J 61(6):9–13
18. Reason JT (1997) Managing the risks of organizational accidents. AP Company and Brookfield Vermont, USA
19. CAP 712 Safety management systems for commercial air transport operations. CAA UK (2002)
20. CAP 760 guidance on the conduct of hazard identification. Risk assessment and the prediction of safety cases for aerodrome operators and air traffic controllers. CAA, UK (2006)
21. Safety Management Systems (2008) Guidance to organizations, Version 2, CAA, UK
22. Safety management systems for flight operations and aircraft maintenance organizations—A guide to implementation, TP 13881, Transport Canada (2002)
23. Risk Management & Decision Making in Civil Aviation, TP-13905B. Transport Canada (2004)
24. State program of safety of flights of aircraft of civil aviation of the Russian Federation (2008) Approved by the decree of the RF Government dated May 6, 2008. No. 641-R. Meeting of the legislation of the Russian Federation, 19 May 2008, no 20, art 2373

25. Versan VT (2003) Actual problems of introduction of the Federal law “On technical regulation” Stand Qual (5):17–21
26. Kulavskiy VG (2011) The management of operational safety of the airline. In: Kulavskiy VG, Sharov VD, Kudryavtsev AG. Problems of safety, no 3, pp 2–48
27. Safety of aircraft: textbook/under the editorship of V. S. Ivanova. M: Izd-vo vvia im. N. E Zhukovsky (2003), 365 p
28. Zubkov BV Safety management. Part I. Program of safety. Guide safety management: a training manual. In: Zubkov BV, Polyakov PM, Karmysov MV.– M.: MSTUCA, 132 p
29. Zubkov BV (2007) Safety and aviation safety. Part II Provision and maintenance of airworthiness of the aircraft: teaching aid. In: Zubkov BV, Sakac RV, Kostikov VA – M.: MSTUCA, 76 p
30. Zubkov BV (2011) Methods of risk assessment in system safety management. In: Zubkov BV, Prozorov SE. Scientific Bulletin of MSTUCA, no 174, pp 7–11
31. Sakach RV (1989) Safety. InL Sacac RV, Zubkov BV, Davidenko MF. –M.: Transport, 239 p
32. Federal aviation rules “Preparation and performance of flights in civil aviation of the Russian Federation”. Order of Ministry of Transport of the Russian Federation 31 July 2009, no 128
33. Hamrakulov IV (1991) Efficiency of use of flight information region. Hamrakulov IV, Zubkov BV—M.: Transport, 175 p
34. Novozhilov GV (2012) Problems of safety of aviation and transport and other complex human-machine systems: monograph. In: Novozhilov GV, Rezchikov AF, Neimark, MS.—M.: Publishing Center “Science”, 200 p
35. Guide to safety management (SMM). Doc. 9859. ICAO, 3rd edn., 300 p
36. Vishnyakov JD (2006) Others development and implementation of normative methodological base of the integral indicators of risk of emergencies. In: Vishnyakov DV et al. Safety Problems and Emergencies, no 1, pp 50–70
37. Orlov AI (2012) Organizational and economic modeling: a textbook in 3 hours Part 3. Statistical methods of data analysis A.I. Orlov—M.: Izd-vo MGTU im. N. Uh. Bauman, 624 p
38. Stupakov VS (2005) Risk-management t:manual. In: Stupakov VS, Tokarenko GS. –M.: finances and statistics, 288 p
39. Risk management: Risk (2000) Sustainable development. Synergy. In: Vladimirov VA, Vorobyov, Sergei Salov YI and others –M.: Nauka, 431c
40. Gnedenko BV, Belyaev YK, Solovyov AD (2013) Mathematical methods of reliability theory. The main characteristics of reliability and statistical analysis. In: Gnedenko BV, Belyayev YK, Solovyev AD. –M.: Publishing House “Librian”, 584 p
41. Kulba VV (2011) The use of scenario and indicator approach for managing survivability, durability and safety of complex technical systems. In: Kulba VV, Kononov DA, , etc. Scientific Edition. –M.: ICS RAS, 116 p
42. Henley E (1984) Reliability of technical systems and risk assessment. In: Henley EJ, Kumamoto H. Per. from English. M.: Mashinostroenie, p 528S
43. International Standard ISO 31000 (2009) Risk management. The principles and guidelines, 1st edn. ISO, 32 p
44. Orlov A (2012) Approaches to General risk theory. In: Orlov AI, Pugach OV. Managing large systems. Issue 40. M.: IPU Russian Academy of Sciences, pp 49–82
45. Kulba VV (2008) Emergency management. In: Kulba VV, Arkhipova NA. –M.: Publishing House. RGGU, 474 p
46. Baldin KV (2006) Risk management: teaching aid. Baldin KV –M.: Eksmo, 368 p
47. Vishnyakov YD (2008) A general theory of risks: proc. A manual for students. proc. institutions. Vishnyakov YD, Radaev NN, 2nd edn. M.: Publishing center “Akademiya”, 368 p
48. Dubois D (1990) The theory of possibilities. Annex to knowledge representation in computer science. Dubois D, Prad A. Per. with FR. –M.: Radio and Communication, 288 p
49. Fedyukin VK (2005) Managing quality of processes. Fedyukin VK. - SPb.: Peter, 202 p
50. Zhivetin VB (2009) The system of civil aviation safety (analysis, forecasting, management). In: Zhiveti VB. vol 17/n. Institute of problems of risk, Information and publishing center “Bon Anni”, Moscow, 370 p

51. GOST R 51901.1-2002 risk management. Risk analysis of technological systems. M.: Publishing House of Standards, 40 p
52. Orlov AI (2012) A new paradigm of applied statistics. In: Orlov AI. Zavodskaya laboratory. Diagnostics of materials, vol 78, no 1, part I, pp 87–93
53. Matveev GN (2010) The method of proactive safety management in aviation enterprises.abstract. dis.....candidate of technical Sciences: 05.22.14/Matveev Georgiy Nikolaevich, 28 p
54. Kirpichev IG (2008) Fundamentals of construction and functional development of the information-analytical system of monitoring of the lifecycle of aircraft components. In: Kirpichev GI, Kuleshov AA, Shapkin VS. –M.: GosNII CA, 287 p
55. Malinetskii GG (2012) Cognitive centers as a tool for safety management of complex systems. In: Malinetskii GG, Mitin N et al. Problems of the safety of complex systems. Proceedings of XX International conference. Moscow, December, pp 3–18
56. Makhutov NA (2001) Features of application of methods of risk analysis of the system "human-machine-environment" based on fuzzy sets. In: Makhutov NA, Kryzhevich, OB etc. Problems of safety in emergencies, no 1, pp 99–110
57. Probabilistic Risk Assessment Procedures Guide for NASA Managers and Practitioners. NASA, Washington, DC, 20546, 2002
58. Kuklev EA (2007) Automated system of monitoring and control of aircraft on the criterion of acceptable risk based on database management. In: Kuklev EA. Scientific bulletin of MSTUCA, no 122(12), pp 37–44
59. Kuklev EA (2003) Evaluation of the level of safety in civil aviation in risky situations based on chains of random events. Sci Technol Transp (2):4–14
60. Shtovba SD (2007) Design of fuzzy systems by means of MATLAB. Shtovba SD. –M.: Hotline—Telecom, 288 p
61. Hodinski, IA (2007) Methods for soft estimation of values: monograph. Hodinski IA. Tomsk State University of control systems and Radioelectronics, 152 p
62. Spesivtseva A (ed) Elements of preventive risk management in the operation of the system of objects. SPb.:Ed. SPb GPU, 132 p
63. Borisov AN (1990) Decision making based on fuzzy models. Examples of use. In: Borisov AN, Kromberg OA, Fedorov IP. Zinatne, Riga, p 184
64. Drozdov AV (1994) Formalization of expert information in the logical-linguistic estimation of complex systems. Drozdov AV, Spesivtsev AV. Technical cybernetics, no 2, pp 89–96
65. Sharov VD (2007) Possibility of fuzzy set theory in assessing risk factors of an aircraft accident. Sharov VD, Guziy AG, Gladkin SM. Problems of safety, no 12, pp 10–18
66. Barzilovich EJ (2006) Optimal control systems condition-based decisions, and proactive adverse situation: monograph. Barzilovich EY, Lonchakov YV, Nikolaikin M.: TEIs, 144 p
67. Viktorov VS (2007) Modeling, and analysis of testability of the on-board aircraft systems. Viktorova VS, Stepanyants AS. Reliability 3(22):62–71
68. Logvin AI (2010) Reliability and technical diagnostics. Part II "Technical diagnostics of REO": a training manual. Logvin AI, Yakovleva DA. MSTU CA, 56 p
69. Improving the technological processes of repairing aircraft. Interuniversity collection of scientific papers. Executive editor, Doctor of Technical Sciences, prof. E.A. Konyaev. –M.: MSTU GA, 84 p
70. Causal model for Air Transport Safety. Final report. NLR, Amsterdam (2009)
71. Lin P-H (2011) Safety management and risk modeling in aviation next generation infrastructures foundation, P.O. Box 5015 2600 GA Delft, The Netherlands
72. Orlov AI (1980) Optimization problems and fuzzy variables. Orlov AI. – M.: Knowledge, 64 p
73. Orlov AI (2006) Theory of decision-making. Orlov AI. –M.: Exam, 573 p
74. Kolmogorov AN (1998) Basic concepts of probability theory, 3rd edn. Kolmogorov AN. FAZIS, Moscow, 144 p
75. Makhutov NA (2006) Development and application of methods of risk management in problems of technogenic safety and technical regulation. Makhutov N. Problems of safety and emergencies, no 1. –M: VINITI, pp 35–49

76. Makhutov NA (2009) Ensuring the safety of critical facilities based on the vulnerabilities. Makhutov N, Petrov VP, Reznikov DO. Safety problems and emergencies –M.: VINITI, no 2, pp 50–69
77. Makhutov NA (2012) The use of risk matrices when performing risk assessment and prioritization of protective measures. In: Makhutov NA, Reznikov DO, Petrov VP, Cuxeva VI. Safety problems and emergencies, no 1, pp 82–92
78. Kravets VA (1980) The method of "fault tree" in the analysis of safety systems to the oil and gas industry. Kravets VA. –M.: Internet Registry, 40 p
79. Shvyrev V (1992) Probabilistic analysis of the safety of nuclear stations. Methods of execution. Shviryaev Y and others. –M.: IAE them. IV Kurchatov, 266 p
80. Hooks SP (2007) Methods of analysis and assessment of risks in the tasks of safety management of complex technical systems. In: Kryukov SP, Bodunov SL, and others. SPb: The Corporation "Aerospace equipment", 460 p
81. Beshelev SD (1980) Mathematical-statistical methods of expert assessments: textbook. Beshelev SD, Gurevich FG. –M.: Statistics, 263 p
82. Orlov AI (2011) Organizational and economic modeling: a tutorial: 3 h. Orlov AI. –M.: Izd-vo MGTU im. N. E Bauman, 2011, Part 2; Expert evaluation, 486 p
83. FSF CFIT checklist evaluate the risk and take action safety digest, vol 15, no 7/8, July–August 1996, pp 26–29
84. McCarthy (1999) U. S. Naval Research Laboratory, D. Schwartz, AT&T. Modeling risk with the flight operation risk assessment system (FORAS). ICAO Conference in Rio-de-Janeiro, Brazil, November 1999
85. GOST R 51898-2002 (2002) Aspects of safety. Rules for inclusion in the standards. M.: Publishing House of Standards, 9 p
86. Article-by-article commentary to the Federal law "On technical regulation" on an evaluation of conformity. Aronov IZ, Rybakov AM, Terkel AL, Stern LM –M.: ZAO "Publishing House" "Economy", OOO "Cero", 176 p

Chapter 2
Analysis of Navigational and Meteorological Risks that Influence Civil Aviation Safety

2.1 Classification of Emergency, Critical, and Failure Situations in Aviation Enterprise

Let us formulate the main principles of the approach to safety in the airlines:

1. **The obligation of the higher executive management**

   - executive management declares the priority of the safety;
   - safety is a significant component of production;
   - executive management controls the safety of the production, as well as all personnel responsible for safety.

2. **Responsibility and discipline all employees**

   - safe work performance is the most important part of the evaluation system management;
   - safe flight operations and ground handling encouraged;
   - upon completion of any work, control;
   - the implementation of the rules and procedures of each responsible officer.

3. **Zero accidents as it is clear the task**

   - the officially declared goal is safety; therefore, it is possible to ensure that the staff understands and supports this goal;
   - establishes a definite system of management and motivation in those areas where it is necessary to focus staff on safety;
   - in respect of all events should be investigated;
   - The purpose of the investigation is always to search for the causes, not culprits.

V. D. Sharov et al., *Probabilistic-Statistical Methods for Risk Assessment in Civil Aviation*, Springer Aerospace Technology, https://doi.org/10.1007/978-981-16-0092-0_2

4. **The audit and quantification activities**

    - guide ensures regular safety audits and the involvement of the necessary specialists in this process;
    - safety assessment is focused on the work control of personnel and production activities;
    - to assess the level of safety are regularly carried out analyses of production results.

The responsibility of the staff

- every airline employee has the authority to discharge their official duties and is responsible for ensuring safety;
- every employee has the opportunity to participate in defining standards and safety procedures;
- all personnel using a system of voluntary reporting,

provides information about the variances and incidents to the management for taking timely preventive measures;

- to ensure the safety of the personnel of all divisions of the airline, they can interact with each other.

The main objectives of safety policy are as follows:

- The orientation of all levels of leadership and management for safety.
- Improve control system safety.
- Precise requirements for duties and responsibilities of the staff.
- Availability of the necessary funds to transfer safety information.
- The involvement of everyone in the safety process.
- Provide the necessary training and maintenance of skills of management by safety.
- Absolute safety is the main objective of the airline.

Responsibility for developing and monitoring the implementation of the safety program is assigned to one of the airline's managers responsible for safety inspections. It is responsible for programming, establishing an internal safety reporting system, investigations, and safety requirements.

To create an airline management system in emergency, crisis, and failure situations, it is necessary to develop a set of regulatory documents regulating various production procedures.

The goals of creating an action plan for the airline's structural divisions in crises can be formulated as follows:

- ensure a quick and organized transition to action in the crisis situations;
- promptly and correctly classify the crisis and choose the appropriate course of action;
- ensure the effective use of all means and resources to minimize damage in crises;

- collect materials to investigate and determine the causes of the crisis and its consequences for the airline;
- the plan regulates the actions of the airline's structural divisions in crises that arise as a result of acts of unlawful interference in the activities of civil aviation, aviation accidents and incidents, actions of natural and human-made factors in peacetime;
- the plan is fundamental for drawing up standard instructions for the actions of the airline's personnel in crises.

To ensure a systematic approach to managing the airline in extreme conditions, we will classify emergency, crisis, and failure situations.

An emergency is an event or incident of a natural or human-made nature that can be accompanied by the death of personnel and passengers, significant material losses, and requires the involvement of a large number of forces and means to prevent its development and eliminate its consequences.

A Crisis Situation (CS) is an event or incident that radically changes the standard working order to a critical one, characterized by a lack of time, the need to make responsible decisions, and a lack of forces and means to implement effective responses at the moment in order to eliminate the threat to human life, damage to property and equipment, or eliminate the onset of such consequences.

The reason for the occurrence of crises can be of various aviation events.

Aviation events—following the Rules of Investigation of Accidents and Incidents (RIAI-98)—are divided into

- aviation accidents, which, depending on their consequences, are divided into the following:
- aviation accidents without loss of life (accidents);
- aviation accidents with human casualties (accidents);
- aviation incidents;
- industrial accidents.

Aviation Accident (ACCID)—an event involving the use of an aircraft that occurs from the moment when a person entered the aircraft with the intent to fly to the moment when all persons who were onboard for flight left the aircraft, and during which a person is injured fatally as a result of being in the aircraft, except in cases where the injuries are caused by natural causes, self-inflicted, or caused by other persons, or when injuries are caused by a stowaway hiding outside areas where passengers and crew members are normally allowed access.

Note. For statistical uniformity only, an injury resulting in death within 30 days of the incident is classified as a fatal injury.

(1) the aircraft is damaged or destroyed structures, resulting in the following:

   - structural strength is violated, and technical or flight characteristics of the aircraft deteriorate;
   - a significant repair or replacement of a damaged item is required.

(2) the aircraft goes missing or ends up in a place where access to it is impossible.

Note. An aircraft is considered missing when its official search has been stopped and the location of the aircraft or its wreckage has not been established. The decision to stop searching for a civil aircraft in distress is made by the Federal civil aviation authority of Russia.

An aviation accident without the loss of human life is an aviation accident that did not result in loss of life or loss of any of the passengers or crew members.

An aviation accident with human victims (catastrophe) is an aviation accident that resulted in the death or disappearance of any of the passengers or crew members; accidents also include the death of any of the persons onboard during their emergency evacuation from the aircraft.

Aviation incident—the event associated with the use of the aircraft, which took place from the moment when a person came onboard with the intention of flight, until all persons aboard to flight, left aircraft, and due to deviations from the normal operation of the aircraft, crew, management services, and support operations, the influence of the external environment that may affect the safety of the flight, but not ended an aircraft accident.

Serious aviation incident—an aviation incident, the circumstances of which indicate that an aviation accident almost occurred. The following features characterize serious aviation incidents:

- the aircraft goes beyond the expected operating conditions;
- the occurrence of significant harmful effects on the crew or passengers (smoke, vapors of caustic substances, toxic gases, high or low temperature, pressure, etc.);
- significant reduction in the performance of crew members;
- significant increase in psychophysiological load on the crew;
- serious bodily injury by any person on the aircraft;
- significant deterioration of stability and handling characteristics and flight or strength characteristics;
- the occurrence of a real possibility of damage to vital elements of the aircraft as a result of an explosion, fire, non-localized destruction of the engine, transmission, etc.;
- destroying or disconnecting controls;
- damage to aircraft components that are not related to an accident.

Illegal interference in aviation activities is defined as illegal actions (inaction) that threaten the safety of aviation activities, which result in accidents with people, material damage, hijacking or hijacking of aircraft, or create a threat of such consequences.

Industrial accidents are divided into:

- aircraft damage on the ground;
- emergency.

Damage to an aircraft on the ground—an event related to the maintenance, storage, and transportation of aircraft, in which the vessel is damaged that does not violate its power elements and does not impair flight characteristics, the elimination of which is possible in operational conditions.

Emergency—an event related to the operation of the aircraft, but not related to the ACCID, in which one of the following consequences occurred:

- death of someone onboard the aircraft as a result of deliberate or careless actions of the victim or other persons, and not related to the operation of the aircraft;
- the death of a person who entered the aircraft without permission and hid outside the areas where access is open to passengers and crew members;
- death of crew members or passengers as a result of adverse environmental effects after an aircraft is forced to land outside the airfield;
- death or fatal injury to any person outside the aircraft as a result of direct contact with the aircraft, its components, or the gas-air jet of the power plant;
- destruction or damage to the aircraft on the ground, resulting in a violation of the strength of its structure or deterioration of flight characteristics as a result of a natural disaster or violation of maintenance technology, storage, or transportation rules;
- hijacking an aircraft on the ground or in flight, or hijacking such a vessel for hijacking.

Failure situation—a situation when there was a failure that led to a violation of the schedule of the aircraft. These include:

- mass (more than 1000) congestion of passengers at the base airport caused by a violation of the regularity of flights for the following reasons:
- failure to provide aircraft for three or more flights at the base airport;
- non-provision of aviation fuel and de-icing liquids for flights at the base airport and three or more terminal and intermediate airports;
- mass (more than 1000) non-arrival of passengers on flights operated from the base airport due to a failure in the operation of urban transport;
- closure of the base airport for technical reasons (rolling the aircraft out of the runway, failure of radio equipment for landing and ATC, the condition of the runway, and others declared by NOTAMs), which will lead to mass congestion of passengers.

Serious bodily injury—a bodily injury received by a person during an aviation accident or incident that caused severe or dangerous harm to health, as well as non-life-threatening harm to health, which is severe in consequences, and which

- requires hospitalization for more than 48 h within seven days of receiving the injury;
- resulted in any bone fracture (except for simple fractures of fingers, toes, or nose), dislocations in large joints of the limbs and vertebrae, or compression of soft tissue with crush syndrome;
- is associated with ruptures of biological tissues that cause severe bleeding, damage to nerve trunks, muscles, or tendons;

- caused damage to any internal organ, as well as penetrating wounds and bruises to the eyeball, accompanied by vision disorders;
- is associated with getting thermal, chemical, and other burns of the second or third degree, regardless of the area of the lesion, or any burns affecting more than 5% of the body surface or upper respiratory tract burns; with exposure to electric current, accompanied by a violation of consciousness, respiratory disorders; as well as frostbite of the third and fourth degree or general cooling of the body;
- is associated with the confirmed fact of exposure to infectious and toxic agents or ionizing radiation;
- resulted in moderate to severe concussion of the brain and spinal cord, and traumatic intracranial hemorrhage;
- resulted in the termination of pregnancy.

Figure 2.1 shows the classification of extreme situations in the airline's operation.

In the following sections of this Chapter, all the requirements and situations listed in this section will be considered in relation to the operation of navigation systems of civil aircraft in the conditions of radio attacks and various meteorological conditions.

2.2 Analysis of Possible Unauthorized Actions on the Navigation System of Civil Aircraft

2.2.1 Classification of Possible Radio-Electronic Attacks on the Navigation System of Civil Aircraft

One of the risk factors affecting the safety of civil aircraft is the markedly increasing activity of various criminal groups. There have been tragic cases of aircraft crashes and an explosion at Domodedovo airport. These cases indicate that civil aviation is the target of terrorist attacks.

At the same time, these cases are characterized by the fact that until recently, terrorist attacks on the aircraft were carried out by a direct "frontal" attack—explosives were carried directly onboard the aircraft. Because the Russian Federation, as well as other countries of the world, pays quite a lot of attention to this problem, we cannot expect a particular increase in threats in this direction, which, of course, does not exclude cases of individual crimes.

If we assume that the criminal elements will not refuse to commit crimes against the aircraft, we can assume that a possible and, most likely, objectively natural, new direction in this activity is associated with the application of **radio attacks on the flight control systems of the aircraft.** In addition, according to the media, criminal groups have the necessary engineering and scientific potential, financial resources, and an appropriate experimental base.

Confirmation of this can be found, for example, in the response of the Deputy General Director –flight Director of JSC "Aeroflot–Russian Airlines" I.P. Chalik

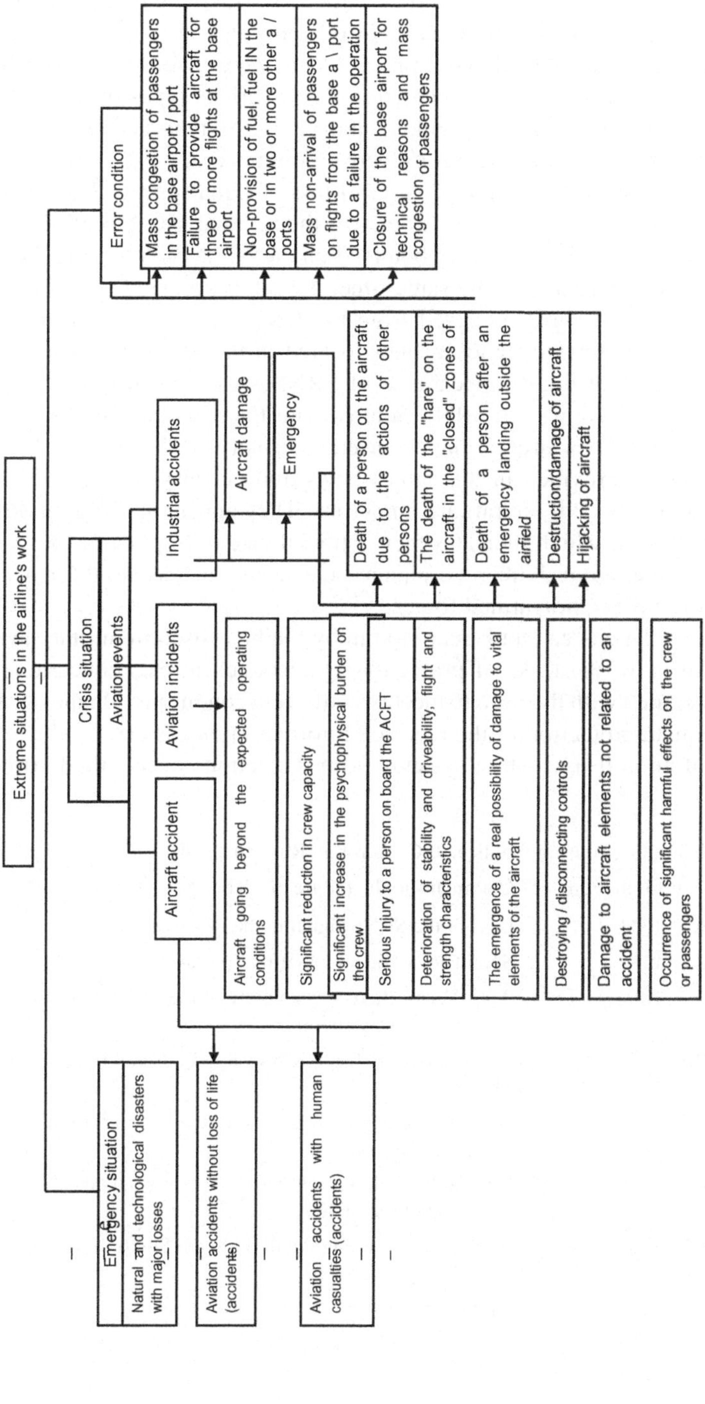

Fig. 2.1 Classification of extreme situations in the work of the airline

from 26.01.2015 to the request of the Deputy head of the airworthiness maintenance Department of the Federal air transport Agency E.V. Isakov, where, in particular, information is provided that only in one (!) week only in one (!) region 7 (!) cases of serious failures in aircraft flight control systems were recorded, caused by the impact on the navigation systems included in them, various types of radio attacks, from sources of unknown origin and unknown location. All recorded failures, if they were not removed, could have the most severe consequences.

Moreover, there were more than 20 cases of loss of signals of the Satellite Radio Navigation System (SRNS), which were noted by the crews of JSC "Aeroflot–Russian Airlines" since 16.01.2015 in the area bounded on the North by Venev, in the South by the border with Georgia. At the same time, the pilots received verbal information about an unstable signal received from the RNC in the Rostov zone, which is confirmed by the crews of other airlines located in the same area. At the same time, it was noted that the disappearance of SRNS signals occurred at any time of the day. Table 2.1 shows the SRNS failures with the date and flight number of Aeroflot–Russian airlines aircraft that occurred in January 2015 [1].

The information provided in this table shows quite clearly that failures in the navigation systems of civil aircraft are numerous. They appear at different times of day and time of year, which indicates their artificial origin. In practice, we can say that we are talking about deliberate, planned electronic attacks on the navigation systems of the civil aviation armed forces.

Parrying such threats cannot be carried out by methods used in military aviation since the solution of the tasks of protection against such threats concerning civil aircraft is associated with the need to meet two different conditions. On the one hand, ensuring adequate protection of the received information necessary for the reliable functioning of aircraft navigation systems from the actions of qualified criminals,

Table 2.1 SRNS failures during flights of JSC "Aeroflot–Russian airlines"

No	Date	Flight number	Comments to the failure of SRNS
1	16.01.2015	1311	Loss of the SRNS signal in the Rostov zone
2	16.01.2015	1183	Spontaneous signal loss after 30 min
3	17.01.2015	1860	After Venev, there was a failure of the SRNS to Georgia, then there was its restoration
4	18.01.2015	1123	When departing from Sochi was a failure of both sets of SRNS
5	18.01.2015	1624	At Simferopol airport, there was a failure of the SRNS with subsequent restoration
6	18.01.2015	1186	During the flight to Volgograd, there was an unstable operation of the SRNS
7	18.01.2015	1284	Failure SRNS during the flight to the South of the city, followed by reduction
8	27.01.2015	1103	Failure of the SRNS in the Rostov zone

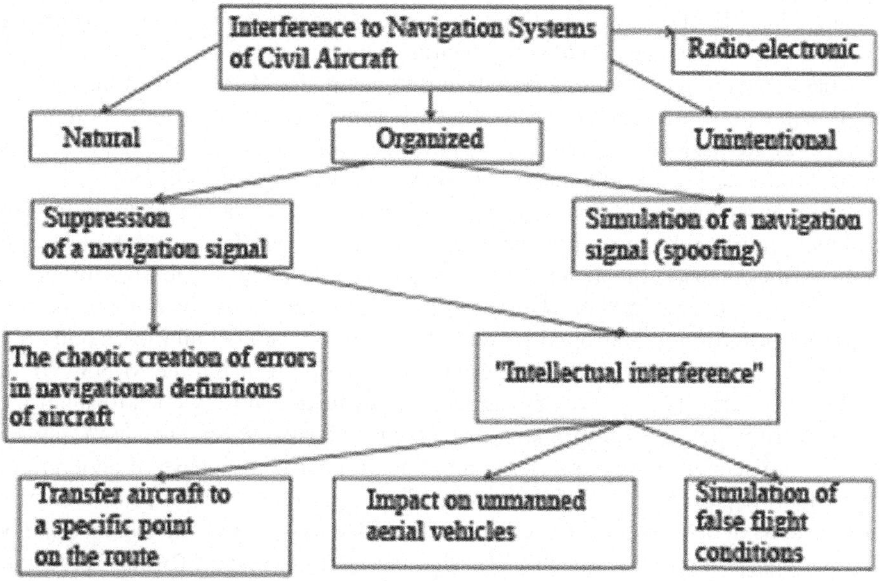

Fig. 2.2 Classification of errors for navigation systems of civil aircraft

and on the other, providing controlled access to a wide range of people in order to inform them about the air situation, which is due to the requirements of ICAO [1].

The classification of interference that, in our opinion, will be most accessible for electronic attacks carried out by criminal groups on the navigation systems of the ACFT of the CA is shown in the block diagram (Fig. 2.2).

Potential interference caused by criminal elements can be conditionally divided into the following types:

- suppression of the navigation signal (destruction of the navigation messages);
- imitation of a navigation signal (spoofing).

It should be assumed that when suppressing a navigation signal criminals, given the vulnerability and complete in safety from radio attacks of navigation systems of civilian aircraft compared to military aircraft, will most likely use narrow-band interference, rather than broadband, requiring a more significant energy resource, especially since all the main tactical and technical characteristics of civil aviation avionics are publicly available. To create the most straightforward narrow-band interference destroying the navigation message, it is enough to have a relatively simple and cheap device, which can lead to disruption in tracking the location of the aircraft or to a sharp decrease in the signal-to-noise ratio.

Imitation interference (spoofing) makes it possible to carry out navigation definitions, but with large errors. When creating such interference, signal transponders installed on Unmanned Aerial Vehicles (UAVs) of the Satellite Radio Navigation Systems (SRNS) can be used. Thus, not only the navigation signal of the SRNS is imitated, but also the fact of the transmission of interference from a moving object. This interference can be divided into the following types [1].

Chaotic creation of jamming signals during navigation definitions of the aircraft. At the same time, criminal groups do not have as their goal to transfer aircraft to the point of airspace they need, but only to wreak havoc in determining their location (to smear the navigation characteristics of the aircraft in space and time).

"Aiming navigation" interference. The purpose of creating such interference by criminal groups is the impact on navigation systems to transfer aircraft to a specific point in the airspace. Such interference becomes the most dangerous during the transition to zonal navigation when the aircraft itself can choose its route in the area of proper provision with ground-based flight facilities. This possibility is quite real since the method of navigational determination of aircraft using the SRNS and the guidance documents on the organization of air traffic are entirely open to any user.

In determining the degree of danger of all of the above interference, it is necessary to take into account the nature and conditions of a particular phase of an aircraft's flight and the particularities of regions in which interference may occur to navigation systems of civil aircraft. For example, "targeted navigation" interference can be diverse, depending on the specific task that the aircraft is currently solving. Situations with the installation of such navigational interference (chaotic or sighting) are considered in sufficient detail in the literature, which deals with the organization, conduct, and actions associated with electronic warfare. Concerning civil aviation, new options for organizing interference effects on the navigation equipment of the ACFT CA appear.

One of the possible types of recent terrorist attacks against civilian aircraft may be the massive use of UAVs as obstacles with the appropriate software and tactics of use. Thus, a real threat to the safety of the ACFT CA can be created.

Radio engineering attacks on the navigation systems of the ACFT CA for various flight conditions can be conditionally divided into two types:

- criminal influences taking into account adverse flight conditions;
- imitation of treacherous flight conditions.

2.2.2 *Possibility of Intentional Interference for the Purpose of the Total or Fragmentary Destruction of Navigation Information*

The low power of the SRNS signal allows the creation of transmitters that suppress consumer receivers. This threat comes from jamming transmitters that criminal groups can use due to their tasks and goals.

According to the studies, possible sources of intentional interference with the SRNS are individual terrorists or "hackers" who are looking for opportunities to create obstacles to the use of the system in any limited area by carrying out terrorist acts. To assess the threat of limited terrorist suppression, an assessment was made of

Table 2.2 Indicative characteristics of a jamming transmitter for terrorist suppression

Power, Watt	Time of action					
	1 h			1 day		
	Cost, $	Weight, pounds	Volume cubic inch	Cost, $	Weight, pounds	Volume cubic inch
10	50	1	50	60	11	250
100	300	10	500	400	110	2500
1000	3000	100	5000	4000	1100	25000

the cost and weight and size characteristics of the noise interference transmitter [2]. Table 2.2 illustrates the relative size of the interference transmitter and overhead. It was noted that the required financial costs are proportional to the output power and depend on the operating time of the transmitter. A 100 W jammer should cost about $ 300 and have the size of a shoebox, while a 1000 W jammer will cost more than $ 3000 and has the dimensions of a small suitcase. The volume and weight of the transmitter increase significantly if the transmitter requires a working time of one day. Based on these data, the study determined that it is realistic to have a 100 W jammer transmitter; using a 1000 W transmitter for terrorist purposes is much less likely due to its high cost.

Risks associated with possible failures or interruptions in the operation of navigation systems of the ACFT CA are reduced to the stability of the reception equipment of consumers of navigation information, which is characterized by a wide variety of options for its construction, resulting from a variety of tactical and technical requirements for consumers, weight, size, energy, and cost restrictions. Various types of equipment are distinguished by the use of open (SPS) and closed (PPS) signals, the number of time and frequency tracking channels with parallel, serial, or multiplex signal processing, and the possibility of combining other navigation systems and tools with the equipment. Consumer Navigation Equipment (CNE) can be integrated with an inertial navigation system, Doppler radar, actual airspeed sensors, barometric altimeters, etc. The most rational pairing is with an Inertial Navigation System (INS), when combined with high accuracy storage of coordinates with high accuracy of their determination.

In general, the risk of an airline operating the navigation systems of an aircraft of civil aviation is determined [2].

- logic and stages of operation of receiving equipment;
- structure of navigation radio signals;
- energy characteristics of navigation signals;
- radio signal processing algorithms;
- the channeling of the consumer receiver;
- methods for integrating navigation information;
- the frequency–time structure of the interference at the input of the receiver.

The direct influence of interference is possible at the stages of searching, introducing into synchronism, and measuring the time delay (tracking).

2.2.3 Energy Analysis of Intentional Interference Possibility in GPS and GLONASS

Analyzing energy ratios and interference balances and main navigation signals, two evaluation criteria can be used at present [1, 2].

- a criterion based on a comparison of the level of interference with a certain threshold level;
- a criterion based on an estimate of the suppression coefficient.

Below are energy ratings for both of the above criteria.

The threshold level analysis is as follows.

The suppression coefficient is entered K_s, which means the minimum necessary ratio of the interference power to the useful signal power at the input of the suppressed receiver within the bandwidth of its linear part, which excludes with a given probability that the receiver will perform a specific function. Numerous studies show that detection prohibition is highly likely to occur with a noise/signal ratio at the receiver output $K_{i\text{- thres–out}} > -10$ dB, failure of support with a probability close to 1, when $K_{i\text{-thres-out}} > 3$ dB.

The threshold level of interference

$$P_{i-thres-in} = P_{sig}\frac{\Delta f_s}{\Delta f_{rec}}K_{SC}K_{IP}K_{i-thres-out},$$

A threshold interference/signal ratio, accordingly,

$$K_{i-thres-in} = \frac{P_{i-thres-in}}{P_{sig}} = \frac{\Delta fs}{\Delta f_{rec}}K_{SC}K_{IP}K_{i-thres-out},$$

where

P_{sig}—*signal at the input of the consumer information receiver*;

$\Delta f_S/\Delta f_{rec}$—the ratio of interference bands to the receiver;

K_{SC}—signal compression ratio;

K_{IP}—coefficient of increasing the threshold level of interference when applying for special interference protection;

$K_{i\text{- thres-}out}$—coefficient of excess interference over the signal at the output of the receiver to achieve a given effect.

The energy potential of the interference source, at which the threshold level of interference is reached, as a function of distance, the expression determines the interference sourced consumer information:

Table 2.3 Relative thresholds for tracking civil signals in the three bands of the GPS system

	Relative power at the receiver input, dB	Relative tracking threshold, dB
L1 C\A	0	0
L2C	−2,3	+0.7 (accompanied by phase capture + 6 dB)
L5	+3,7	6,7

$$P_{IS}G_{IS} = \frac{P_{i-thres-in}(4\pi R_i)^2}{G_{rec}\lambda^2}\frac{1}{\beta\gamma}$$

where P_{IS} G_{IS}—energy potential of the interference source;

G_{rec}—the antenna gain of the receiver of the consumer information;

R_I—the distance between the source of interference and the consumer information;

B—the relative level of reception of interference and signal by the antenna of the consumer information receiver;

γ—coefficient that takes into account the difference between the interference polarization and the receiving antenna of the consumer information receiver.

Absolute threshold interference values are determined by the signal level at the antenna input, which for GPS is −164.5 dBW (with a transmitter power of 26 W), and for GLONASS is −165.5 dBW (with a transmitter power of 15 W).

Relative thresholds for tracking civil signals in the three ranges of the GPS system are presented in Table 2.3.

The twofold gain in the relative threshold for the L2C and L5 signals (signal-to-noise ratio at the receiver input) is determined by improving the structure and processing of the signal.

From a comparison of the threshold levels of the two systems, taking into account the structure of their construction, we can conclude the following:

- each GPS channel is individually more noise-resistant than a single GLONASS channel since for individual channels the threshold levels of detection and measurement of the GPS are higher than the threshold levels of the GLONASS channels, which to some extent is explained by the difference in the signals used and their processing;
- the GLONASS system is more comprehensive than the GPS system since the first one can potentially use information from other frequency channels not suppressed by interference when the threshold level is exceeded in some frequency channels;
- the integral “system” threshold level of interference, defined as the total interference power that interferes with the performance of the system as a whole, for the GPS system is equal to the threshold level of a single channel; for the GLONASS system, the “system” threshold level of interference exceeds the threshold level of a single channel by several frequency channels times, that is, 12 times, that is, more than an order of magnitude.

The suppression range of the navigation receiver, at which there is a prohibition (difficulty) in capturing or disruption (increase in errors) of track, is determined by its spatial position relative to the interference source, its energy potential, and the structure of the interference used by the hardware interference protection methods.

The combination of the last three factors determines the suppression threshold, which is a parameter for the dependence of the potential of the interference source $P_{is}G_{is}$ and relative range R_{rec} of consumer information and interference source (Fig. 2.3). The relationship of the relay coefficient is also presented here K_r*G_{is} interference equipment and relative range R, at which the navigation receiver is suppressed.

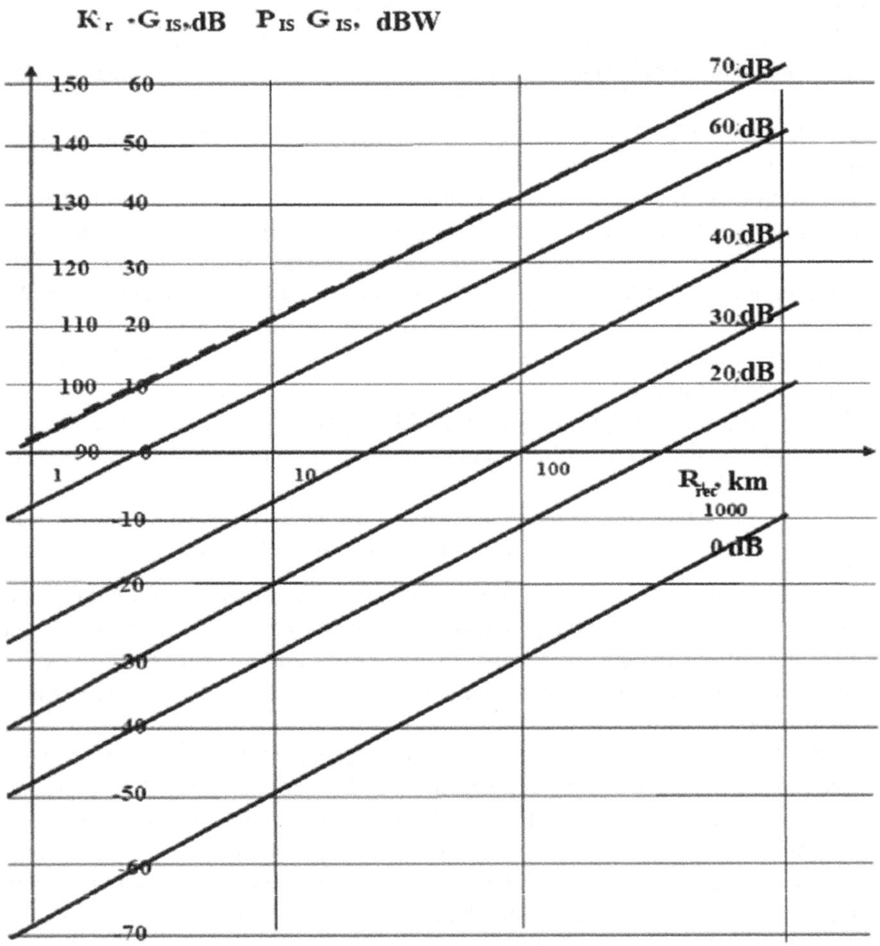

Fig. 2.3 Dependence of energy P_{is} G_{is} potential and the retransmission coefficient K_r*$G_{is,}$ source of interference from $R_{rec,}$ when the threshold ratio is set, interference/signal

Table 2.4 Line of sight D_{ls}, km

h\H, meters	30	500	1000
10	35,4	104,6	142,5
500	114,1	183,3	221,2
1500	181,1	250,3	288

The effect of the interference source on the receiver appears only if the relative range does not exceed the line of sight, which is presented in Table 2.4 depending on the height of the interference source h and the height of the receiver carrier H. The calculation was carried out according to the formula $D_{ls} = 4{,}1(\sqrt{h} + \sqrt{H})$.

If the source of interference is located on the ground with a height of 10–15 m, the radius of potential suppression of GPS receivers of ground and low-altitude objects will be 35–40 km, and aviation objects with a flight height of 500–1000 m −100–150 km. Interference sources placed on air carriers with an application height of 1000–1500 m will lead to suppression of navigation receivers used at low altitudes (<30 m) at a line-of-sight range of 120–180 km.

To illustrate, consider the placement of an interference source with an energy potential of 10 dBW on the ground. In this case, suppression of receivers of low-altitude carriers within a radius of 35—40 km will occur at threshold levels of 50 dB and lower; at a flight altitude of 500–1000 km, suppression in a radius of 100–150 km will occur at threshold levels of 35 dB and lower. It can be seen that these levels significantly (from 10 to 40 dB) exceed the threshold levels set for both GPS receivers and GLONASS receivers. Thus, civilian receivers without special jamming schemes can be suppressed by a 10-watt jammer, at least in a radius of 150 km, both in search and capture mode and in signal tracking mode.

The greatest danger to receivers is represented by simulated interference transmitters, which can be built according to a relay or generator circuit [2].

The construction of equipment according to the relay scheme is unlikely due to the high technical requirements for the relay. Indeed, in order to achieve a zero threshold corresponding to an effective simulation noise at a distance of about 10 km, it is necessary to realize the gain of the relay path (~ 110 dB). This is because the interference is attenuated along the path from the source of interference to the input of the receiver $K_{atten} = \frac{1}{4\pi R_c^2} S_{ANT}$ раз. Here, R_c—range of counteraction and $S_{ant.}$—effective area of the information consumer's receiving antenna with S_{ANT} equal to 100 sm^2 and $K_{atten} = -111$ dB. If the gain reaches 120 dB, the region corresponding to the threshold of 0 dB expands to 30 km (Fig. 2.3, dashed line). This threshold is adopted from the condition of signal similarity and interference. At the same time, the problem of removing the system from synchronism is not solved, which will require addressing noise interference and, accordingly, to a station with high potential. Besides, and this is fundamental, the relay signal at the input of the navigation receiver is always late relative to the direct one, which excludes its use when the GPS receiver is in tracking mode by a signal with a P-code.

Table 2.5 Absolute threshold noise levels at the receiver input, dBW

Channel	GPS L1	GPS L1, L2C	GLONASS L1	GLONASS L1, L2	GPS + GLONASS C
Interference					
Noise Sighting	−140,5	−136,8	−135	−132,7	−132
Harmonic	−150,5	−146,8	−138	−134,7	−134
Simulation generator room	−161,5	−157,8	−162,5	−148,2	−148

The construction of a generator transmitter of imitation interference to civilian receivers is quite possible, since the generator of the pseudo-random sequence polynomial is known, and it can be easily reproduced in jamming equipment.

The increase in the reliability, accuracy, and noise immunity of the GPS and GLONASS global navigation satellite systems are expected to be achieved by applying a pure signal in the L2 band and integrating both systems, which is also equivalent to increasing the number of frequency channels of the generalized system. This leads to an increase in the total absolute threshold level of interference at the input of the integrated navigation receiver, which is reflected in Table 2.5 for three types of interference: noise impact, harmonic, and simulation generator.

Notes

1. Thresholds are calculated for the tracking phase and signal levels at the receiver input:1. Thresholds are calculated for the tracking phase and signal levels at the receiver input:
   GPSL1–164.5 dBW, L2C–163.8 dBW;
   GLONASS L1–165.5 dBW, L2–162 dBW.
2. The threshold ratio for simulated interference is 3 dB.
3. For GLONASS, a "systemic" threshold level has been adopted.

Table 2.6 for various uses of the navigation space system channels shows the hazardous ranges of organized interference sources calculated based on the absolute threshold levels presented in Table 2.5, depending on the operating mode and power of the interference transmitter, taking into account the attenuation of the latter due to the spatial polarization and additional frequency–time signal processing.

From Table 2.6, in comparison with Table 2.4, it follows that the aggregation of the navigation channels and systems:

- significantly reduces the maximum range of the suppression of the navigation receivers (from 2 to 4 times);
- makes the gang or to use the apparatus with the formation of a multifrequency spectrum, making it difficult for her technically and raising the price, or we can go to defensive interference in the frequency band of the integrated navigation system (~400 MHz), a leak will cause to increase the threshold level of 17 dB and lower-range suppression in 7 times (at equal power with the sighting of an obstacle);

Table 2.6 Range suppression of navigation receivers, km

Channel		GPS L1	GPS L1, L2C	GLONASS L1	GLONASS L1,L2	GPS + GLONASS C
Interference	W					
Noise targeting	0,1 1,0 10,0 100,0	85 280 850 2800	70 230 700 2300	60 200 600 2000	40 132 400 1320	38 125 380 1250
Harmonic	0,1 1,0 10,0 100,0	280 829 2800 8290	230 736 2300 7360	87 278 870 2780	49 157 490 1570	46 147 460 1740
Simulation generator	0,1 1,0 10,0 100,0	900 2880 9000 28800	740 2368 7400 23680	391 1251 3921 12510	252 803 2520 8030	236 755 2360 7550

- provides an increase in the acceptable threshold level of interference for 8 to 16 dB, which is below the required two orders of magnitude;
- does not solve the main fundamental task of ensuring acceptable threshold levels of signal to background interference that is significantly distant light sources (40 miles further) with a real frequency and power characteristics within.

Analysis by the valuation method of the suppression ratio is as follows [2].

Electronic vehicle navigation system implementing the location of the aircraft via the satellite radio navigation system can be affected; the fact of the suppression of noise receptor consumer information from the satellites is determined by the excess power interference above the level of the signal at a certain threshold-reduction factor, the value of which is determined by the capacity of the interference range to be suppressed, receiver parameters, and characteristics of the interfering signal.

Figure 2.4 shows a diagram of the relative positions of the satellite, the consumer receiver, and the interference source.

Fig. 2.4 Diagram of the mutual location of the satellite (EST), consumer receiver, and Interference Source (IS)

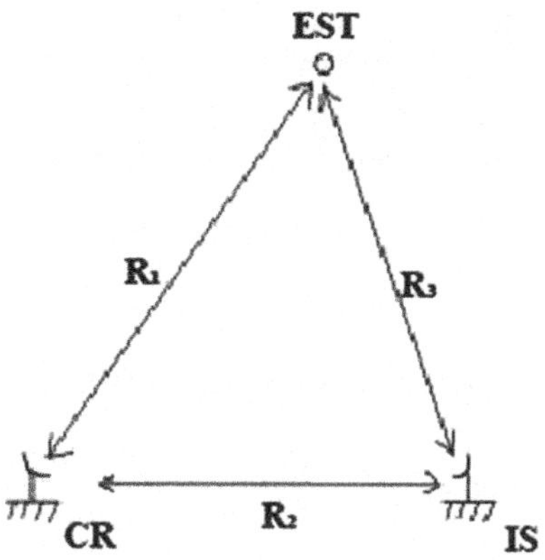

The signal level received by the consumer receiver has a value defined by the expression:

$$P_{cr} = \frac{P_0 G_0 G_{cr} \lambda^2 \Delta f_0 K_0 \delta_0}{(4\pi)^2 R_1^2}, \tag{2.1}$$

where

P_0G_0—energy potential of the artificial Earth Satellite Transmitter (EST), with radiation power P_0 and antenna gain G_0;

G_{cr}—the gain of the receiver user;

λ—wavelength of the radiation signal;

Δf_0—width of the radiation signal spectrum;

R_1_distance from the EST to the consumer receiver;

K_0—coefficient of polarization losses;

δ_0—loss of signal power in the atmosphere on the track R_1.

The level of the interference signal received by the consumer receiver is defined as

$$P_{IS} = \frac{P_{IS} G_{IS} G^1_{CR} \lambda^2 \Delta f_I K_p \delta_L}{(4\pi)^2 R_2^2} \tag{2.2}$$

where:

G^1_{CR}—a gain of the side lobe of the consumer receiver in the direction of the interference source;

$P_{IS}G_{IS}$—energy potential of the interference source;

Δf_I—width of the interference spectrum;

R_2—the distance from the user receiver to the source of the interference;

K_P—coefficient of polarization losses;

δ_L—power loss interference in the atmosphere on the road R_2.

Considering that the state of the atmosphere determines the central signal power loss and for a useful satellite signal is determined by the value of the atmosphere height of 30–40 km, it can be considered with a sufficient degree of accuracy that $K_0 \approx K_P$ is the gain of the side lobe of the consumer receiver in the direction to the interference source.

The suppression coefficient of the consumer receiver is determined by the ratio of the power of the interference source and the power of the useful signal at the input of the consumer receiver from (2.1) and (2.2) as

$$q = \frac{P_{IS} G_{IS} G^1_{CR} \Delta f_\Pi R_1^2 K_p}{P_0 G_0 G_{CR} \Delta f_0 R_2^2 K_0} \tag{2.3}$$

If the interfering signal falls into the main lobe of the consumer receiver, then

$$G^1_{IS} = G_{IS}$$

For satellite radio navigation systems, malfunctions and the issuance of reliable information occur at a suppression coefficient $q \geq 1$.

Correlation-spectral processing at the receiver of the consumer of satellite signals provides an improvement in the signal-to-noise ratio, which is determined by the ratio of the spectral width of the input signal (Δf_0) to the spectral width (Δf_{proc}) at the output of the processing device.

For civilian C/A GPS signals, which equally applies to GLONASS and GALILEO, the values of the signal-to-interference gain are determined as

$$\Delta f_I = n_0 \Delta f_{proc}, \, n_0 = \frac{\Delta f_0}{\Delta f_{proc}} \tag{2.4}$$

Provided that at the input of the consumer's receiver, there is an interference of any incoherent type: pulse, quasi-simulation, noise, then the gain for interference is determined as

$$q = \frac{P_{IS} G_{IS} G^1_{CR} R^2_1 K_p}{P_{IS} G_{IS} G_{CR} R^2_2 K_0 \sqrt{n_0}}, \tag{2.5}$$

$$\Delta f_{IS} = \Delta f_{ob} \sqrt{n}, \quad n = \sqrt{\frac{\Delta f_{IS}}{\Delta f_{ob}}} \quad \text{when } \Delta f_{IS} \leq \Delta f_0 \tag{2.6}$$

Substituting (2.4) and (2.5) into (2.3), we obtain the expression for the suppression coefficient of the GPS consumer receiver at $\Delta f_I = \Delta f_0$.

For the case if $\Delta f_I > \Delta f_0$, there is an energy loss coefficient due to the width of the interference spectrum:

$$m = \frac{\Delta f_{IS}}{\Delta f_0}, \tag{2.7}$$

then the suppression coefficient will be determined as

$$q = \frac{P_{IS} G_{IS} G^1_{CR} R^2_1 K_p}{P_0 G_0 G_{CR} R^2_2 K_0 m \sqrt{n}}, \tag{2.8}$$

When the suppression coefficient q is equal to or greater than unity, which ensures a high probability of inoperability in the normal mode of the consumer receiver, the value of the energy potential of the interference source is determined as:

$$P_{IS} G_{IS} \geq \frac{P_0 G_0 G_{CR} R^2_2 K_0 m \sqrt{n}}{G^1_{CR} R^2_1 K_p} \tag{2.9}$$

For a GPS system, the characteristics are:

- energy potential $P_0G_0 = 240–750$ W;
- orbit height $R_1 = 20{,}000$ km;
- polarization loss coefficient $K_0 = 0{,}5$, determined by the mismatch of the circular polarization of the satellite transmitter and the linear polarization of the consumer receiver;
- for C/A signals $\Delta f_0 \approx 1{,}5$ MHz, for P/Y signal $\Delta f_0 \approx 20$ MHz, while the spectrum of the signal at the input of the processing device for these signals $\Delta f_{proc} \approx 0{,}25$ Hz, therefore the gain for C/A signal $n_{C/A} = 1{,}5*10^6/0{,}25 = 6*10^6$ and for P/Y signal $n_{P/Y} = 2*10^7/0{,}25 = 8*10^7$;
- the level of the side lobes in the standard placement of the consumer receiver on an airplane is G10 = 0.03 (−15 dB), for the viewing angles to the source of interference;
- polarization loss coefficient $K_L \approx 1{,}0$, taking into account the influence of noise along the side lobes on the main and cross-polarization.

The graphs of Fig. 2.5 show the dependence of the energy potential level of the interference source depending on the distance to the consumer's receiver for C/A and P/Y GPS signals, if the GPS signal spectra and the interference source are combined.

The need to simultaneously suppress multiple frequency channels of the navigation system, as noted above (formula 2.7), will require increasing the energy potential of the interference source by a factor equal to the ratio of the total emission spectrum of satellite or satellite transmitters to the spectrum of one channel (formula 2.9).

Therefore, for the GLONASS system, which uses the frequency selection of each satellite, the number of satellites must increase the energy potential of the interference

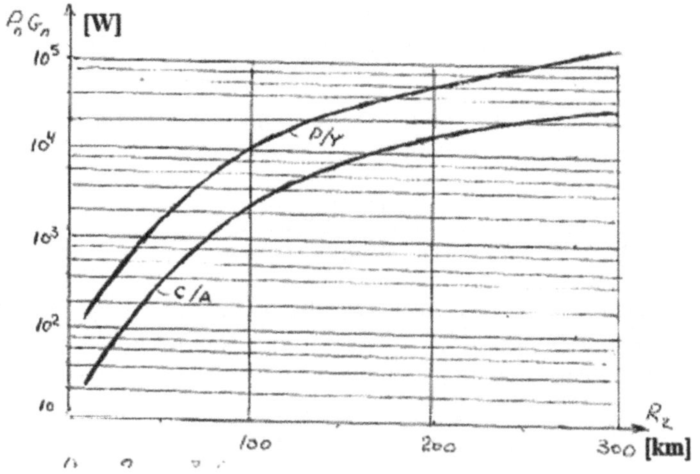

Fig. 2.5 Dependence of the energy potential of the interference source on the distance to the consumer receiver for C/A and P/Y GPS signals

source by 24 times in the system. For the GPS system, the number of times is proportional to the frequency band of the used channels L1, L2 to L5.

If deliberate artificial interference should provide the formation of barrage interference in the entire frequency band used by space navigation systems, then the value of the energy potential should be 250 times for C/A signals, and 25 times more for a P/Y signal than for a signal with the conjugate spectrum for one channel, that is, for obstructive C/A and P/Y signals, the energy potential of the interference signals is almost the same.

The possibility of reducing the energy potential of the interference source is the use of coherent relay interference. In this case, the correlation-spectral processing of the interference signal will be with less loss.

The suppression coefficient (2.8) will be determined by the following equation:

$$q = \frac{P_{IS}G_{IS}G_{CR}^{1}R_{1}^{2}K_{p}}{P_{0}G_{0}G_{CR}R_{2}^{2}K_{0}m}n^{\frac{1}{x}-1} \tag{2.10}$$

where the exponent x should be from 1 to 2, determined by the quality of the coherence of the interference signal compared to the useful signal. With complete identity $x = 1$, and with noise incoherent interference $x = 2$.

Thus, with coherent relay interference, it is possible to reduce the energy potential of the interference source by n times, that is, by two to three orders of magnitude.

2.2.4 *System Vulnerability Analysis for Navigation Systems of Civil Aircraft*

The vulnerability of navigation systems of the aircraft of civil aviation to the terrorist threat of violation of their work is determined by two groups of factors [2]:

- electronic suppression of the GNSS signals at the input of the navigation receivers of users;
- physical defeat (or capture) of ground objects navigation satellite system centers and other facilities control and monitoring stations laying side information from the Navigation Spacecraft (NSC), communication nodes, differential control and correction stations of consumers, etc.

With the terrorist threat, we can accept the rules of interference with certain restrictions on the types of noise, their power and duration, and methods of their production. These restrictions are determined by the capabilities of terrorist groups in the part weights and dimensions of jamming equipment.

This image of the possible threats leads to the fact that civilian users of GNSS impossible to create within their technical capabilities and equipment cost-effective electronic counter-interference and to parry such threats should focus on the specific system and procedural solutions of the transport complex in such conditions.

It should also be noted that it is possible, though unlikely, terrorists use enough high-energy suppression, such as tools that are installed on vehicles or trailers. Under these conditions, counteracting the effect of the use of such powerful tools of terrorist suppression should also be systematic or procedural methods.

It should be noted that in addition to the classification of consumers according to vulnerability to interference are also an essential stage of the transport operation, in which the suppression system.

For civil aircraft, the most critical stage is the approach stage and the final element is the landing of the aircraft.

In the short to medium term, we can take the next most likely appearance and main characteristics of the station of the terrorist suppression of GNSS.

The main inputs to the station of the terrorist suppression GNSS are as follows:

- the purpose of the station is to provide suppression of GNSS in critical areas of transport operations (at airports, on unequipped sites, etc.);
- the location of the station ground in the area is quite close to the estimated zone of suppression. The height of the base is within the architectural building height of the zone;
- suppressed navigation signal L1 civil signal GPS L1 or GLONASS;
- the effective required range of suppression—up to 30 km;
- power Autonomous, time of jamming up to 1 h;
- volume and weight must match the weight and dimensions of the model passenger Luggage (suitcase, bag weighing up to 30 kg).

Based on such inputs by the experts were identified the likely technical characteristics of the station and its elements:

- method of suppressing noise associated spectrum interference;
- radiated transmitter power of 50 watts (efficiency to the capacity of the primary power 40%, weight 15 kg, a volume of 20 DM);
- antenna horn with a useful beam width 60×30 degree (gain 10 dB, weight 5 kg, a volume of 10 DM);
- consumed from the batteries 200 W of power, the battery power consumption 0.2 kW/HR (when $U = 12$ W about 20 ampere-hours). Battery weight 8–10 kg, a volume of 20 DM;
- the corresponding cooling system of the transmitter (convection, evaporation, heat pipes), allowing heat dissipation and ensuring the autonomous operation of the transmitter within the specified time.

For vulnerability assessments and measures to enhance the stability of aircraft using satellite navigation, it is fashionable to apply some numerical and experimental techniques, the main of which are the following:

- mathematical modeling;
- Hil simulation (the combination of a mathematical model with the real system hardware);
- an experimental study with mathematical imitation of those elements that is impossible for whatever reason to implement in the real world.

The Hil simulation is based on the fact that the magnitude of the interference signal calculated for a specific tactical situation is fed to a controlled attenuator, and then also summed with the computed navigational signal SRNS for this situation, and turned into analog form. From the combiner, the signal is input from the real receiver SRNS and further evaluated the effect of the interference on this receiver.

A full-scale experimental study is because the aircraft served a real signal. Not to interfere with navigation tools not participating in the experiment, a disturbance is served onboard test of the device on frequencies other than frequencies of the SRNS. Onboard, this passed signal is converted in frequency SRNS, served on a controllable attenuator which takes into account such elements of the tactical situation, as side lobes of the radiation Pattern of the Antenna (PA), the phasing of signals and others, and summed at the receiver input with real signals SRNS. Further, evaluate the impact on receiver interference in real conditions as close to real.

The results of mathematical modeling showed that after a certain averaging, they are close to the results of direct energy calculations of the effect of interference on the GNSS receiver; therefore, a further calculation method of determining the fact of suppression is used in a further stability analysis.

Earlier in this section, energy criteria for the effects of interference on a user's receiver were evaluated. To assess the vulnerability of systems, it is necessary to move from such criteria to criteria related to the main parameters of safety systems. The main such criterion may be the **criterion of the probability of the completion of a flight under the influence of interference**. This very complex multi-parameter criterion requires taking into account the tactical situation with a massive number of parameters and source data. Therefore, to assess the impact of interference in the case of terrorist suppression, we can use a more simplified **criterion for assessing the magnitude of the distance to the point of completion of the flight**. For example, for approaches, this point may be the line of transition to the visual piloting method. Further, taking into account the fact that the point of completion of an operation for terrorist suppression, as a rule, is close to the location of the jammer, we can take the suppression range as a system criterion.

The maximum permissible value of this range should ultimately be determined for each type of operation based on modeling. Also, the fact of the presence in the terminal area of technical means of detecting an interference suppressor is of great importance for solving the problem of the possibility of completion or termination of the flight.

2.3 Analysis of Meteorological Risks in Aviation Enterprise Operation

Adverse weather conditions can cause severe accidents and even plane crashes. In addition to the risks relating to the safety of aircraft CA, upon the occurrence of such phenomena, it is necessary to consider the risks associated with the inefficient use of the temporary and material resource.

Here is an example. Helicopters are used, including changing the watches on the drill stations located on the coast, in remote areas, and even in the seas or ocean. In the event of adverse meteorological conditions is the ban on flights. While such ban is in effect until the occurrence of favorable meteorological conditions. Thus, to change the operation of one watch to another in this period. However, for a simple watch that continues to be on drilling, we have to pay higher (sometimes twice) wages, and the watch, which arrived for work, also needs to pay for downtime.

The main task of the Meteorological Support (MS) flight is the preparation and production of measurement and forecast information to notify pilots and other aviation personnel on the conditions of weather that may affect the safety of aircraft operations. That is, the meteorological information is an integral component of system-wide information management today and tomorrow, consisting of aeronautical information, along with data on flights, traffic flows, and so on [3, 4].

Currently, in the field of meteorological support in the Russian Federation, processes are underway aimed at harmonizing the domestic weather support system with foreign ones. The following is a brief analysis of meteorological elements and weather conditions, the absence (or distortion) of information about which can lead to air accidents.

2.3.1 Air Temperature

Inaccurate information about the actual *temperature* and its temporal and spatial changes can lead to incorrect accounting of aircraft performance characteristics during dispatch takeoff and landing control services. The higher the temperature, the greater the takeoff and run length, the speed of separation, and landing of the aircraft, as well as changes in the maximum permissible altitude of the aircraft. Fuel consumption depends on temperature [3].

Given the relationship of air temperature with other characteristics of the state of the atmosphere—pressure, density, and humidity—the inaccuracy of the air temperature will lead to incorrect data on its density, since the density is not measured, but calculated based on temperature. Information on the actual air temperature in the surface layer of the atmosphere is given in local reports of actual weather at the aerodrome, on surface weather maps. The air temperature at heights is indicated on the maps of absolute baric topography, the map of the tropopause. In local reports of actual weather at the aerodrome, air temperature is indicated with an accuracy of

1 °C. On the surface weather map in synoptic terms (00, 03, 06, 09, 12, 15, 18, 21 h), air temperature is applied with an accuracy of 0.1 °C, which proves the importance of knowing the actual temperature.

2.3.2 *Air Humidity*

When analyzing the meteorological conditions of the flight, it is necessary to take into account *changes in* air humidity in time and space; the effect of air humidity on the flight performance of aircraft; the relationship of air humidity with its temperature, pressure and density; and the connection of air humidity with weather phenomena that affect aircraft flights.

When providing meteorological support for flights, information is used about the dew point, dew point deficit, and relative air humidity. On surface weather maps, the dew point is indicated with a resolution of up to 0.1 °C. In local actual weather reports at the airport, the dew point is given in whole degrees Celsius. Dew point deficiency is the difference between the air temperature and the dew point. The smaller the dew point deficit, the closer the air to its saturation state, the higher the likelihood of fog. The dew point deficit is indicated on the maps of absolute baric topography. The relative humidity is given in local weather reports at the aerodrome. The above requirements for the accuracy of humidity measurements, as well as in the case of temperature, show the importance of knowing the actual humidity [3].

2.3.3 *Atmosphere Pressure*

When providing meteorological support for flights, when analyzing and assessing weather conditions, five air pressure values are used, which are designated as follows: QFE, QNH, QNE, QFF, and P min.

Pressure QFE is atmospheric pressure at the aerodrome, corresponding to the level of the runway threshold of the landing-operating course. The QFE is given in local actual weather reports at the aerodrome in ATIS transmissions, reported by the air traffic controller to the crew onboard the aircraft before boarding to set up the altimeter. This pressure is indicated in mm mercury column, rounded down to integer values. In METAR telegrams, after the abbreviation RMK, it is given not only in mm mercury column, but also in hectopascals.

Pressure QNH is air pressure at the aerodrome, reduced to sea level under the conditions of a Standard Atmosphere (SA). The cast is carried out using a barometric formula. The QNH pressure is indicated in METAR telegrams in hectopascals, rounded to the nearest integer value downstream of the letter Q. For example, Q1000 means that QNH is 1000 hPa. In international practice, QNH is used to set the altimeter, and in METAR telegrams can be indicated in inches of mercury column after the letter A. For example, A2992 means the QNH is 29.92 inches mercury column. One inch is approximately 25 mm. If the aircraft is located on the runway and QNH pressure is set on the altimeter scale, then the altimeter should show with sufficient accuracy the height of the airfield relative to sea level. If QFE pressure is set, the altimeter should indicate zero altitudes. QNE pressure is the standard sea level pressure of 760 mmHg mercury column. (1013 hPa or 29.92 inches mercury column).

QFF pressure is determined by SYNOPTIC stations and is plotted on surface weather maps in the hPa to 0.1 hPa. QFF is the pressure at the station, reduced to sea level, not in SA, and on the actual weather conditions, taking into account the measured values of temperature and humidity. Data about atmospheric pressure is given to sea level, so they can be used to analyze and forecast the weather on the ground. On the globe, QFF changes from 1085 hPa in the Siberian anticyclone to 877 hPa in the tropical cyclone. Most often, it varies from 950 to 1040 hPa.

Reduced pressure (Rmin.Pref.) is reported in the forecasts on the areas in the service flying at low altitudes [4]. This pressure in the practice of meteorological support of flights is determined by the ground-level weather maps and is given in mm Hg. article rounded to whole numbers less. Atmospheric pressure is the most sensitive meteorological variable, at least for two reasons: first, it is the primary method of determining the height of the aircraft (barometric method), and second, by changes of atmospheric pressure in the time and space to judge about the changes of the weather.

In aviation, to describe the air pressure changes with an altitude of the pressure level used is the reciprocal of the vertical pressure gradient. (Pressure level—the height to which to rise and the pressure has decreased by one). When the temperature increases by 1 °C, the pressure level increased by 0.4% at constant pressure. In the warm air at the same pressure of the pressure level more than in cold air. The pressure stage is used to calculate safe altitudes.

2.3.4 Wind

The wind is blowing the plane off the course that plane is flying. The degree of rotation depends on the relative position of the flight direction and the wind direction, as well as on the flight speed and wind speed. The direction and speed of the wind at the flight level are part of the initial data necessary for a preliminary navigational calculation. Especially crucial for air navigation is the consideration of spatial and temporal variability of the wind. Information about the wind at the flight level is also used to enter into automatic navigation devices and onboard computers.

The plane takes off and always lands against the wind. During takeoff and landing, in this case, the path length decreases. During takeoff and landing with a fair wind, the takeoff and mileage increase. That is, incorrect information about the actual direction of the wind can lead to an accident and even disaster. The plane may simply not have enough runway length.

Sidewind has an adverse and sometimes dangerous effect on the aircraft. To ensure safety for each type of aircraft, the maximum permissible values of the cross-wind speed (its side component), tailwind speed, and even headwind speed are set. Distortion of valid wind data can lead to landing outside the runway [4].

A danger for the flight represents only an extreme wind with a speed of 25–30 m/s. Typically such a wind is gusty when the wind speed varies from 10–12 m/s to 25–30 m/s. Such changes in wind speed lead to abrupt changes in the lift force, which is equally as dangerous during takeoff and landing.

Ground wind at SYNOPTIC stations is determined by the height of 8 m from the surface of the earth, and at the airport on the heights of 6–10 m from the runway. The accuracy of surface wind measurement at aerodromes in speed to 1 m/s direction up to 10°. In ATIS broadcasts for arriving and departing aircraft are given the wind at heights of 100 m and a circle in order to be able to assess the changing wind in the vertical.

When the meteorological support of flights used by the meteorological wind direction is expressed in tens of degrees and is determined by the azimuth of the point of the horizon whence the wind blows. Meteorological wind differs in the direction from the navigation of the wind by 180°. At ground level, the weather in METAR telegrams, given the actual correct wind direction, measured from the geographical (true) Meridian. The period of averaging of the wind is 10 min. In weather on the airfield, compiled code TAF indicates a right prognostic wind. Local weather reports by the airport in ATIS, for takeoff and landing of the aircraft, are given by the magnetic wind, averaged over 2 min. The direction of the magnetic wind is measured from the North direction of the magnetic Meridian. Magnetic wind differs from the right wind if the magnetic declination is 5° or more. When determining the magnetic wind, if the magnetic declination is positive (East), it is subtracted from the right wind direction, and if negative (Western) magnetic declination is added.

2.3.5 Wind Shifts

Characteristics of the spatial variability of the wind are *wind shifts*—a change of direction and/or wind speed in the space, including ascending and descending air currents. Strong wind shifts are considered dangerous for aviation weather phenomena. Distinguish between *horizontal* or *vertical wind shifts.* The basic distance for which the measured horizontal shifts of the wind is 600 m. The vertical wind shift describes the change of the wind vector with height. For practical purposes, meteorological support of flights of the values of the wind shift calculated for a layer thickness of 30 m indicates in m/s at 30 m. Sometimes the wind shift determines the layer thickness of 100 m, and in international documents and foreign literature, there are other units of measurement of wind shift (knots, ft/s, miles/h, etc.). The impact of vertical wind shear on the flight is much more dangerous than horizontal. In almost all airfields, there is the problem in determining (measuring) the shifts of the wind.

Table 2.7 Intensity wind shifts criteria

The intensity of the wind shifts (quality term)	Impact on aircraft management	Vertical wind shifts, m/s at 30 m	Horizontal wind shifts, m/s at 600 m
Weak	Insignificant	0	2
Moderate	Significant	2	4
Strong	Essential difficulty	4	6
Very strong	Dangerous	>6	

In the case of vertical wind, shear landing is possible, or in the beginning of the runway, or even before the start of the runway. During takeoff climb into the zone, the stronger the wind [3, 4]. This means that the lifting force of the aircraft increases with height faster than estimated, and with a strong wind shift, the plane can get on supercritical angles of attack. In turn, this leads to stall, stall the aircraft on a wing and a possible collision of the sun with the earth. If the wind speed with height is not increasing and is decreasing, the picture is reversed. Therefore, often the shifts of the wind at the airport are not determined; however, always predicted. The intensity of the wind shifts are divided into weak, moderate, strong, and very strong. Intensity wind shift criteria are shown in Table 2.7.

2.3.6 *Air Turbulence*

Among the meteorological phenomena that have a dangerous effect on aircraft operations, *atmospheric turbulence* occupies a leading position. Turbulence causes intense aircraft chatter. At the same time [3, 4],

- altitude, course, and flight speed change;
- the stability and controllability of the aircraft, as well as flight comfort, are deteriorating;
- the wear of individual units and units of the aircraft increases.

On takeoff and landing, the chatter is dangerous in that, due to strong turbulence, significant throws of the aircraft up and down from the calculated flight path are possible. Throws up can cause the aircraft to be at supercritical angles of attack, which is most dangerous when the plane takes off. Throws down can lead to a collision with the earth or water surface, which is equally dangerous both during takeoff and during landing. Following the ICAO rules, if the increment of the overload $|An| < 0.5$, then the chatter is weak. Vertical gusts up to 10 m/s correspond to it. At $0.5 < |An| < 1.0$ bump is considered moderate, and vertical gusts in this case are 10–15 m/s. In those cases when $|An| > 1.0$, the chatter is recorded as strong and vertical gusts of more than 15 m/ s correspond to it. The most dangerous type of turbulence for aviation is Clear Sky Turbulence (CST), since it always turns out to be sudden for the crew. Important circumstance should be noted: difficult and sometimes impossible conditions for flying during a thunderstorm are created not only in thunderclouds but also near these clouds.

2.3.7 Clouded Sky and Limited Visibility

The difficulty of piloting an aircraft in the clouds or with poor visibility is the lack of visual orientation and piloting can only be performed using instruments. When flying in clouds or in a zone of poor visibility, occur more often: severe turbulence, icing of aircraft, or other dangerous weather events than when flying outside of clouds, and mirages and colored haze are possible, which make flight very difficult.

Under certain weather conditions, a mirage, a "light pole", arises from the illuminator turned on, and the pilot has the feeling that a "strip" has appeared in front of him. When flying in any condition on an airplane, sidelights are always on. When flying in clouds that have a variable density scatter this light, and a colored haze of variable density is created. Turning off navigation lights when landing is prohibited. Failure to report low cloud cover and reduced visibility to the aircraft can also lead to an accident.

2.3.8 Weather Minimums

The weather minimum is determined by two values —the height of the lower boundary of the clouds (the height of the decision) and visibility on the runway. To ensure the safety and regularity of flights, the following weather minimums are established: *the minimum of the airfield, aircraft, aircraft commander, and type of aerial work* [3, 4].

Decision Altitude (DA)—the established relative height at which the maneuver of approach to the second circle should be started if, before reaching this height, the aircraft commander has not established the necessary visual contact with the reference points to continue the approach, as well as if the position the aircraft in space or its motion parameters do not provide a safe landing. Violation of the minimums or DA can lead to an accident or disaster.

In order to ensure the safety and efficiency of flights in severe weather conditions, the so-called categorized minimums, or **ICAO minimums**, are established. These lows fall into three categories:

- the first category—the height of the lower boundary of the clouds 60 m, and visibility on the runway is 800 m;
- the second category—the height of the lower boundary of the clouds is less than 60 m, but not less than 30 m, and visibility on the runway is less than 800 m but not less than 400 m;

- the third category—the height of the lower boundary of the clouds is less than 30 m, and visibility on the runway is less than 400 m.

The minimums of the third category include three varieties of varying degrees of complexity. The minimum of category III-A provides visibility on the runway of at least 200 m, category III-B—not less than 50 m, and category III-C visibility on the runway can be 0 (zero) m.

2.3.9 Visibility Range

In civil aviation, two characteristics of visibility at the surface of the earth are used: visibility itself and visibility on the runway, which means the distance at which the pilot of an aircraft located on the centerline of the runway sees the marking of the runway or lights indicating its contours from the cockpit centerline [3]. For the object to be visible, it is necessary that the illumination of it and the background is not less than a particular value, and that there is a brightness contrast between the background and the object. Brightness contrast can vary from zero to one. The minimum brightness contrast that distinguishes the human eye is called the threshold of contrast sensitivity and is designated B. For daytime conditions, B = 0.02, and for nighttime, B = 0.6–0.7. In order to ensure safety when determining visibility in the interests of aviation, ICAO set B = 0.05 for daytime conditions. All meteorological processes and phenomena that contribute to increasing the brightness of the atmosphere layer reduce visibility. Such phenomena include precipitation, fog, snowstorms, dust storms, etc.

A special place among all the characteristics of landing visibility is visibility, which is understood as a very long distance along the glide path, which when degraded visibility, the pilot from the cockpit landing of the aircraft may be on the threshold of perception to detect or to identify the beginning of the runway or its associated primary signal lights [4]. Of all the known characteristics of visibility is determined only by the value of meteorological visibility, that is visibility only in the horizontal direction at the height of the observer's eye (approximately 1.5 m), which in the practical work of the pilot is not used. The connection between the landing and the meteorological range of visibility is quite tricky. Landing visibility depends on the height and structure-cloud haze, the transparency of the atmosphere in the terminal phase of the glide path, and light and photometric performance runway. It is considered that with the height of the lower border of the cloud 300 m and below landing meteorological visibility is less, but with more high cloud cover, they are virtually the same. Thus, not making available to pilots and other aviation personnel information about conditions, weather, visibility, can lead to an accident.

The main meteorological phenomena, visibility should include the precipitation (continuous rain, precipitation, drizzle, or ice crystals, rain, snow, sleet, snow pellets, hail, drifting), blizzards, fog, dust or sand storm, mist, and haze. Flying in the area specified weather hazards is dangerous not only because of degraded visibility but also because of the possible icing up to 500–1000 m, the snowdrifts, and turbulence in the lower atmosphere.

2.3.10 Aircraft Icing

Icing is one of the most adverse meteorological phenomena, which significantly affects the safety and regularity of flights of aircraft and helicopters [4]. Severe icing may cause an accident. The deterioration of aircraft performance when flying in an icing zone depends on the intensity of icing, the amount of ice deposited on the surface of the aircraft, the shape of ice deposits, and the structure of ice. All types of aircraft are subject to icing, including supersonic aircraft, since any supersonic aircraft flies at a relatively low subsonic speed during takeoff and landing. Certain regularities have been established that allow us to predict the icing of the aircraft in flight. In the zone of atmospheric fronts, icing is more common, and its intensity is much higher than in the case of intramass clouds. On warm fronts, icing is most often observed in warm air in the temperature range from −10 to −20 °C. The more active the front, the more all the frontal characteristics, and signs can be traced in the front zone and the more intense the icing in the warm front zone. On cold fronts, as well as on warm ones, icing is more often observed in warm air. The intensity of icing depends on the type of cold front. On cold fronts of the first kind, the icing intensity usually does not exceed moderate (if there is cumulonimbus cloud cover on this front), and on cold fronts of the second kind, the icing intensity is usually strong. On occlusion fronts, icing is most often observed in the area of the occlusion point in the area of the cold front. The temperature layer in which icing occurs most frequently on all fronts remains approximately the same: from −5 to −20 °C. If there is an intramass cloud cover in the flight area, icing of the aircraft is also possible. Its intensity depends on the cloud shape and its water content. Of all intramass cloudiness is often observed in layered icing and cumulonimbus clouds. In layered clouds, moderate icing prevails in intensity, while in cumulonimbus clouds, moderate or severe icing prevails. From performing flights in conditions of possible icing for aviation, layered rain clouds for flights of mainline aircraft and layered clouds for flights of helicopters and small aircraft represent the greatest danger. Thus, deliberate misrepresentation of data about the SYNOPTIC situation may lead to an incorrect forecast of possible icing of the aircraft.

2.3.11 Glazed Frost and Ice-Covered Ground

Glazed is one of the most dangerous meteorological phenomena for flights, which can seriously complicate the activities of aviation. When there is ice, problems arise when preparing aircraft for flight, and airfields to receive and release aircraft and helicopters. If ice is deposited on the surface of an aircraft on the ground, takeoff of such a vessel is prohibited. If ice covers the runway or taxiways, it significantly reduces the friction of the wheels of the chassis on the concrete, which makes it difficult to run up and run the aircraft. Landing on an icy runway with a crosswind is a danger of the aircraft evading the desired direction of movement and rolling out of the runway [4].

The ice-covered ground is a weather phenomenon (not precipitation) that occurs when, under the influence of solar heat, snow and ice on runways, taxiways, and aircraft Parking areas first melt and then freeze when it gets cold, forming a layer of ice on the surface. The presence of "spots" of ice, sleet, and stagnant water on artificial runways is particularly dangerous for takeoff and landing of aircraft and helicopters. In this case, water and snow drawn into the running engine can disable it, and ice, snow, and even water in the lane make the braking process poorly controlled.

2.3.12 Thunderstorms, Squalls, Whirlwind, Rain Showers, Hail

Thunderstorms are one of the most dangerous weather phenomena for aviation. Cause the danger of thunderstorms [3].

- intense turbulence in the clouds, which can cause severe turbulence and overload exceeding the maximum permissible values;
- severe icing at altitudes where the air temperature is below 0 °C;
- the possibility of hitting the aircraft with lightning;
- heavy rain.

Of all the meteorological factors associated with thunderstorms, the most dangerous for flights is atmospheric turbulence.

Tornadoes and squalls often accompany thunderstorms.

Whirlwind is the most destructive atmospheric phenomenon. Its vertical length can reach 12–15 km (in southern latitudes up to 20 km), the horizontal length reaches several tens of kilometers, and the speed of air movement in the vortex of a tornado (in fact, the wind) reaches values of 200–300 m/s. Squalls, like tornadoes, are a severe threat to aviation. If the plane in flight meets with a tornado, then an accident is inevitable. Rainstorms are natural companions of thunderstorms. Heavy rain is hazardous for flights, as it disrupts the regular operation of aircraft engines, significantly reduces visibility, and complicates the activities of airlines.

Hail also presents an exceptional flight hazard. When we encounter hail in flight, due to the impact of hailstones on the surface of the aircraft, various damages can occur—from dents on the surface to depressurization of the cabin and passenger compartment.

A brief analysis of the vulnerability of data on the values of meteorological elements and certain weather conditions has shown that false information about weather elements and phenomena in the area of the airfield and on the roads, violation of the categorized minima, can lead to accidents or even catastrophes of specific aircraft, as well as to serious violations in aviation activities. The most dangerous and protective weather elements in this regard are air temperature and wind parameters at ground level and at altitudes, atmospheric pressure, cloud cover, and visibility characteristics on the runway, and dew point deficit.

References

1. Zatuchny DA (2018) Ensuring navigation safety of civil aviation in the conditions of possible unauthorized impacts. Inform Analyt Magaz "Radio-Electron Technol", 3(18):96–100
2. Zavalishin OI, Zatuchny DA (2019) Methods for improving the noise immunity of navigation systems of civil aircraft. Inform Analyt Magaz "Radio-Electron Technol 2(21):70–74
3. Bogatkin OG (2009) Fundamentals of aviation meteorology. RGGMU Publishing house, Saint Petersburg, p 339
4. Bogatkin OG (2010) Aviation weather forecasts. Saint Petersburg: Publishing House "BHV-Petersburg", 284 p

Chapter 3
Risk Assessment Methods for the Aviation Enterprise

3.1 Safety Risk Management in Aviation Enterprise as a Task of Situational Management

The well-known concept of risk management is shown in Fig. 3.1.

We can see that risk management comes down to making decisions on the allocation of resources. Accordingly, the risk management methodology, as a set of management methods, is a decision support tool.

The decision-making task allows for different approaches. As noted by academician Pospelov "Many mathematical models can correspond to the same typical problems. Each decision-making model can be implemented not by one but by many algorithms. Moreover, each algorithm can be implemented by many computer programs" [1, p. 6].

It seems that the task of managing the risk of S in an airline can be attributed to the tasks of situational management, which described Pospelov [2]. That task has some properties characteristic of these problems, namely:

1. The uniqueness of the object of management must take into account all the individual characteristics of each airline, the nature and volume of work performed, route network, the fleet of aircraft, the level of "safety culture", etc. For a full description of the airline as a control object, require high costs. As experience has shown, the number of equations in the system can be several thousand.
2. The complexity of formalization of the purpose of existence. On the one hand, the company's goal is to profit from the air, and on the other, each airline declares in its policy of ensuring safety as the primary goal of the activities. This is a known issue "*production*"–"*rotection*", which is given much attention in the SMM ICAO [3]. Also, cannot be formalized all the knowledge since the descriptions do not reflect all the major parameters. For example, it is difficult to consider the level of crew resource management (CRM).

V. D. Sharov et al., *Probabilistic-Statistical Methods for Risk Assessment in Civil Aviation*, Springer Aerospace Technology, https://doi.org/10.1007/978-981-16-0092-0_3

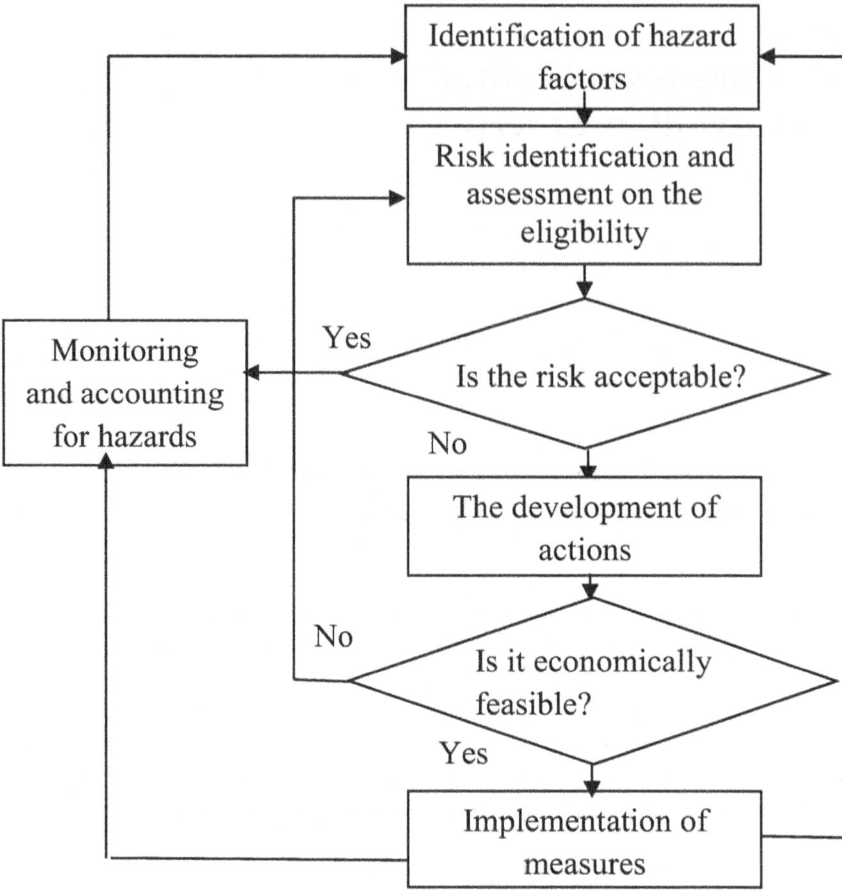

Fig. 3.1 The general concept of risk management

3. Classification of situations, combining them into classes, and the actual risk evaluation mostly subjective, though, with the development of specific management practices, every effort is made to reduce the degree of subjectivity.
4. The dynamic control system must be adaptive, ready to change its functioning. The risk management system at the enterprise is continually improving and evolving.
5. The lack of optimality is impossible to assess the quality of the decision if alternative possibilities have not been tested. If, after the adoption of a decision on risk minimization occurs, the aviation event with small damage, it is evident that when making any other decision, this event would not have occurred or, on the contrary, there would be more challenging.

As noted in the work of Leveson [4], reduced ability to study problems through experience.

Therefore, it is appropriate to talk not about **optimality, but only about the feasibility of a solution.**

The principal conclusion: the S risk management system, as a system of situational management, cannot optimize the management process itself, and S risk management methods remain mostly heuristic.

As noted by a well-known specialist in S from the Airbus J. Nisula concern, risk management will always be subjective; the task is to minimize the level of subjectivity [5].

Currently, various hazardous industries use many methods to manage various risks. S risk management has its specifics; therefore, to develop a standard methodology, it is necessary to analyze the applicability of various methods in the practice of an airline.

3.2 Study of the Known Risk Assessment Method's Applicability to the Aviation Industry

3.2.1 Classification of General Risk Assessment Methods. Applicability Analysis for Qualitative and Combined Methods

There are many different methods of risk management and the different principles of their classification.

In GOST [6] is the description of method 31. GOST [7] proposed seven primary and ten auxiliary methods. The names and descriptions of some of the methods are so different from [6], which is difficult to establish their identity. Two methods from [7] ("Paired comparison" and "Analysis of hidden processes") are not found among the 31 methods of [6].

Stick to classification methods [8], but to put in brackets the names and indices of the appropriate methods of GOST [6].

In [8] is proposed to divide all methods on phenomenological, deterministic, and probabilistic.

The phenomenological method is based on determining the possibility or impossibility of the occurrence of the emergency process and gives the result if we can confidently determine the status of the components of the system. It can be used to determine the relative levels of safety for different types of industrial installations, but not suitable for the analysis of processes, so in practice, airlines rarely used.

The deterministic method involves the analysis of a sequence of accidents, starting from the first event. The process is studied and predicted using calculations and mathematical modeling. Disadvantages are the potential to lose sight of the critical chain of events and the difficulty of building adequate models. It can be used in risk analysis for S, given the increasing possibilities of computer technology.

The probabilistic method provides an assessment of the probability of occurrence of the accident and the calculation of the relative probabilities of one or another path of the development process. It is implemented in the well-known, and widely used probabilistic safety analysis (PSA) described in several papers, for example [9–11].

In practice, airlines used a combination of all three methods.

Methods are divided into qualitative, mixed, and quantitative. They can be deductive or inductive.

Methods of expert estimations: Brainstorming (B.01), **Delphi Technique** (B.03). At this stage of development of risk management, including the risk of PD, the role of expert estimates is exceptionally high. Only an expert is capable of processing heterogeneous information based on their experience and intuition and to synthesize an appropriate conclusion [12–14].

Expert evaluations are used in cases when formal methods are too complicated and the original database is insufficient to obtain an analytical solution. Used both group and individual assessment.

Areas of application can be:

- scores and probabilities of events and the severity of their consequences;
- forecasting of event occurrences;
- report on the work of other professionals or organizations.

Apply several varieties of expert assessments, which are determined mostly by the form of the expert survey. The main types are brainstorming, structured or semi-structured interviews, and Delphi method.

The monograph under the editorship of Novikov [15] describes in detail a method of integrated risk assessment, developed in ISP RAS. The method consists of 12 stages and is mostly based on expert assessments.

It should be noted that the activities of the expert in each area should be governed by relevant regulations and implemented in accordance with various teaching materials. An example is a Federal law No. 174-FZ "On ecological expertise". Because the management of S no regulations and guidelines according to the expert prediction, the airlines develop their methods. A detailed analysis of the airlines used in the methods with examples of usage performed by us in [16]. Expert assessments in this book are used in the development of all three variants of the systems of risk management.

Checklists (B.04) **and Structured What-If Technique (SWIFT)** (B.09) are qualitative methods, based mainly on the study of compliance of operating conditions with safety requirements. These methods are widely used in safety management systems. For example, the Canadian manual [17] contains a particular Q-850 program in which checklists are tools for assessing hazards and risk reduction options. Next, a step-by-step scenario of the event and the adoption of corrective actions is compiled.

Checklist methods include fixing defects during audits and inspections, for example, during the SAFA audit. What if method declared as one of the main in the FAA program of the USA [18].

The same group of methods includes the decision-making system when flying to "high-risk airfields" at Volga-Dnepr Airlines.

Hazard and Operability (HAZOP) (B.06) **is** based on the fact that deviations from the usually observed levels of various process parameters indicate the presence of existing or developing problems. The basics of the HAZOP procedure similar to the procedures of control charts. As applied to technical systems [8] is the deviation of a specific variable from the nominal value for the complex human–machine system, which is the airline's deviations, discrepancies, and events that are detectable in the assessment of operational activities.

We can say that this is the primary way of identifying hazards in the practice of airlines [5, 19, 20], including operating helicopters [21]. The method used in the automated system of risk management.

Adjacent to this method: the **Failure Mode and Effects Analysis (FMEA)** (B.13); **Failure Mode Effects and Criticality Analysis (FMECA); Causal analysis** (B.17). All these methods involve the consideration of each component of the system on the subject that brought her from a state of regular operation and as an "abnormality" in this part affects the system as a whole. In [18], the combination of these methods is called analysis of operational activities (Operations Analysis).

Provides for the classification of failures according to the criteria "probability–seriousness". This classification can be specified for each object, the technical devices of the system (in our case—for each of the activities of the airlines), taking into account its specificity. Thus, we come to the "risk matrix" (B28 Matrix of consequences and probability), which in several guidelines [19, 22], and provides a quantitative assessment. Therefore, these methods can be called "mixed" or "semi-quantitative".

In [8] and other sources, for example, in [9], rightly pointed out and the main drawback of these methods is the difficulty or impossibility of their use for analysis of combinations of events leading to the accident.

Preliminary Hazard Analysis (PHA) (B.05) is an inductive method designed to identify hazards at all stages of the system. The results can be presented in different ways, for example, in the form of tables or tree-like system.

The method is recommended in several guidelines. The leadership of the pilots Association [19] provides a technique of drawing up of model "a preliminary list of hazards" (*Preliminary Hazard List*), apply some of the mentioned methods. The source data (*input*) is all the available information about S: aviation events specific to the airline threat external influences recorded human error, etc. Then brainstorm compiled a list of the main dangers in the activities of the airline (*output*). An example of such a list, which contains more than 50 major hazards, for example, a collision of aircraft in the air, CFIT, weather hazards, etc., required to develop a script using the "what if?".

An example of a system of hazard identification-program "INDICATE" the Australian transport safety bureau (ATSB) [23].

To a certain extent, a preliminary analysis of the factors runs continuously in the activities of the airline. The complete system view is presented in SMS airline "Aviashelf" [24].

The effectiveness of this method will be determined primarily by the completeness of dangerous situations. The method involves carrying out a large volume of work and constant updates. The method is most useful for a small airline. For an airline with a large fleet of different aircraft, vast geography of flights, an excellent Charter program, the compilation of a list of dangerous situations will be a significantly more voluminous and time-consuming task.

An inventory of hazards possible based on the classification of factors from aviation accidents states automated system of maintaining the S of the Russian Federation [25] and based on the structure of airline operations, adopted by the IATA [26].

Human Reliability Assessment (HRA) (B.20) is used for the qualitative assessment of events related to human error ("the Method of analysis reliability of personnel"). The method is vital for the human–machine system, which is an airline.

Error staff is an action that is performed or not performed under certain conditions [8]. In the probabilistic safety analysis (PSA), this question is considered from the standpoint of the analysis of personnel reliability [9, 11]. A qualitative reliability analysis is conducted to determine the logico-temporal structure of the algorithm (instructions) to perform each function of the staff under the specific conditions of the work management system and the consequences of possible erroneous actions of personnel.

An example of a method for the qualitative assessment of human errors can serve as a "Maintenance Error Decision Aid" (MEDA) developed by Boeing [27].

Quantitative characteristics of human error are obtained via the **Technique for Human Error Rate Prediction (THERP)**, **Accident Sequence Evaluation Program (ASEP)** or **Human Factors Analysis and Classification System (HFACS)**, a description is given in the work of the Institute of transport safety of the Netherlands (NLR) [28].

The method used in SMS of major airlines: Aeroflot and Utair, which is reflected in [29, 30].

General details of the classification of logical and probabilistic methods are given in [8]. Note the logical-probabilistic method of analysis of Professor I. A. Ryabinin and methods based on different "trees".

The method of Professor I. A. Ryabinin is based on [31] the logical and probabilistic safety theory (LPST), which was developed by him. This theory is based on the logical representation of the development of dangerous situations and mathematical methods for calculating the truth of the functions of the algebra of logic, representing the hazard functions of systems.

The advantage of LPST is its operability in the absence of initial probabilities of initiating events, which, as a rule, is a fundamental difficulty in quantifying the danger of rare events due to the lack of frequency stability in initiating conditions. The method can be applied in the practice of an airline, provided it is adapted, specialized personnel are trained, and specialized software products are used [32]. An example of the practical use of the method for the synthesis of the function of a priori estimation of the S level by the group of factors "Environment" is given in [33].

Logical-probabilistic can be attributed to the method proposed in [34]. The method is based on a scheme for the development of a particular situation from complications of flight conditions to an accident. Similar probabilistic-statistical models are considered in [35].

Methods of Fault Tree Analysis (FTA) (B.14) **and Event Tree Analysis (ETA)** (B.15) are effective methods of the analysis of conditions of occurrence of adverse events.

The fault tree analysis is a method for the identification and analysis of factors that can contribute to the studied undesirable event (called end event). With the deduction of the investigated factors identifier, to build them logically, and present the diagram in a tree view that displays these factors and their logical relationship with the target event. The tree is built from the top down—from the end event to the initiator event, and further to the factors causing them.

The event tree is built in the opposite direction—from initiating event through the intermediate and allows you to calculate risk in terms of the probability of the main event. The method of event tree is the primary tool of the probabilistic safety analysis [9, 36].

In the practice of the airlines, these methods separately are usually used sparingly, only for the evaluation of private events. More efficiently using a combination of both methods, which in [6] has received the name of B16 Analysis of causes and consequences. The paper begins with a consideration of the final event and analyzes its effects with the use of logic elements YES/NO method (fault tree). Then the causes of the conditions or failures are analyzed using the fault tree. International experience of the development of similar systems (I-Risk, ARAMIS, WORM) is presented in [28].

The closest scope is a Causal model for air transport safety (CATS) [37], where you dwell.

The CATS program was developed by a team of scientists of the Netherlands Institute of transport safety (NLR) at the request of the Ministry of Transport of the Netherlands in the period from 2003 to 2009, using best practices and software products universities and research centers of many countries. The purpose is to identify the sequence of events leading to ACCID, the evaluation of the existing "barriers of safety", risk assessment (as ACCID probability and severity of consequences), development of preventive measures.

A list of event types (33 types), which can occur at various stages of the flight. For each event type on each flight stage by the developed causal model. The CATS model is a diagram of a sequence of events (ESD), and fault tree (FT) combined single Bayesian network-based trust model (BBN). A fault tree was built based on the description of aircraft accidents and incidents related to the categories of the ACCID.

CATS used probability density distributions of hazard factors rather than point estimates to estimate the likelihood of an accident. To obtain distributions, various databases were used, the main of which:

- ICAO **Aviation Events Database (ADREP)**;
- DB data of airlines and airports.

Thus, the model used the method from [6] No. 26 Bayesian analysis and Bayesian networks.

The disadvantages of the model:

- the verification of the accuracy of the calculations has not been performed, a simple comparison is supposed with similar models of assessment;
- the model is designed for the safety of air transport in general, the features of individual airlines are not considered, respectively, the real data on airline operation (PI, data on failures) are not involved;
- risk in CATS does not include the value of damage in terms of value, only deaths and loss of aircraft are considered.

Causal analysis (B.17) adjoins tree methods

It is often more convenient when the final event is considered the safe completion of an operation. As an example of the use of the method in airline practice, in [16] we presented an analysis of the chassis non-release using the "Rat Bag" scheme of British Airways.

For events that are more complex can be applied to the cause-and-effect diagram Japanese theorist of management Kaoru Ishikawa "fish skeleton" described in [38]. An example of its use is given in [16]; it is also known that the method used in the airline "Aviashelf" [24].

Levels Of Protection Analysis (LOPA) (B.18) is vital for the airline. It is a mixed method of assessing the risk of an undesirable event or scenario. It is aimed at analyzing the adequacy of management actions or risk reduction based on the selection of pairs of causes and effects and identify the levels of protection that can prevent the cause leading to the undesired consequence. To determine the adequacy of the mitigation of risk to an acceptable level is necessary to calculate the consequences.

This method was used in the development of SMS in the airline "Meridian" in the risk assessment.

Following the methods of [8] do not have a precise match for 31 methods in GOST [6].

Risk analysis using the graph type "graph" is the second (after the tree of events) chart type causality. There are two types of graphs [8]:

- graphs of transitions and States;
- streaming graphs.

Later in [8], it is noted that models of the type "graph" in the form of influence diagrams have long been used in models of safety research. Their dignity is in the comfort of the transition to symbolic models and their basic mathematical formulas of dependence between the previously selected quantitative indicators of safety and the basic parameters of human–machine systems. The analytical expressions can be used for a priori (prior) and a posteriori (actual) assessment of the level of risk.

Process models using graphs are found quite often in S research papers; however, this method has not yet been sufficiently demanded in practical techniques for managing safety risk.

3.2.2 Applicability Analysis for Quantitative Risk Assessment Methods

For the practical needs of the objective quantification of risk is a crucial objective. Unfortunately, at present, a single methodology of such assessment no. In varying degrees, it can be used to calculate the various indicators. The general classification of these methods is given in [8]. Quantitative methods are divided into deterministic, statistical, deterministic-statistical, probabilistic, statistical-probabilistic, deterministic and probabilistic, probabilistic logic, fuzzy sets, bifurcation, expert, etc.

Part of the quantitative methods naturally intersects with qualitative and mixed. Let us consider briefly the characteristics of those quantitative methods that can be applied in the airline.

Deterministic methods applied for assessment of the risks specific technosphere objects with known values of such parameters as the time to reach a hazardous condition, the stock of the resource, the coefficients of the gravity of a given degree of damage. These parameters are calculated as the conditional probability of reaching limit state and economic damages for the most severe accidents. Close to deterministic methods of reliability assessment used in the system of continued airworthiness.

Statistical and deterministically statistical methods for risk assessment of S are based on the compilation of statistics on the incidence (frequency) of occurrence of aviation events and corresponding damages. Based on statistical data of past years by means of known methods of probability theory and mathematical statistics are based on trends and estimates of expected losses, with a significant amount of statistical material for the decision of practical tasks (e.g., tasks of the insurance) probabilistic distribution of events. On this basis, the problem of quantitative risk assessment is being solved.

Probabilistic risk assessment methods are the most complete and rigorous from a mathematical point of view. However, their use presupposes the existence of known probability distribution on the set of events, which in the practice of aviation enterprises is difficult to obtain.

Quantitative methods include methods such as probabilistic logic, methods based on TNM. These methods allow us to obtain numerical values of probabilities of events if we have the original data.

Methods of forecasting the state of aviation technology (analytical, probabilistic, and statistical) have been described in detail [39]. If the full capabilities of these methods have not yet been implemented for the solution of specific problems of risk management of S, they apply.

An important method is the frequency analysis of emergency events. Its purpose is to evaluate the intensity of the implementations of each of the predictable accidents. Unlike the likelihood, the intensity of random events measured in units of inverse time.

Frequency analysis is one of the main stages of risk analysis and involves the following steps [8]:

- a finding of intensities (probability) of accidents;
- identify the events that most significantly affect the intensity (probability) of accidents;
- development of recommendations on the reduction of their intensity.

In the description of the stages of the frequency analysis of the concept of "intensity" is used as a synonym for the concept of "frequency" and "probability". In practice, these three concepts to be synonymous, and this method is used to estimate the probability of manifestation of risk factors.

Methods according to GOST [6], allowing to perform a quantitative risk assessment as most essential to address the objectives of this study are summarized in Table 3.1, which also lists examples of the use of these methods in industry practice.

Other quantitative methods such as nonlinear dynamics methods, methods of analogies, mathematical simulation modeling, indexes, and dangers have not yet found full application in the practice of risk management of S. However, research on the use of these methods in the SMS are carried out. For example, we can assume that in [40] laid the theoretical foundations for the application of the method indexes the dangers and fuzzy assessments for the risk management of aviation activities.

The analysis also showed that legal documents, scientific studies, and practical guidance on risk management use different wording and descriptions of the same methods.

A single classification of methods of risk management currently is also missing.

3.2.3 Limitations of ICAO Recommended a Conceptual Risk Assessment Model

For risk assessment, ICAO proposes to use a matrix, which according to GOST [6], is a matrix of "consequences and probabilities". It is borrowed from the technical manuals on SMS of some states. The first matrices in CA, for example, in CAP 712 [41], had 4×4 size, then, for example 4×5, in [42], later all came to a single matrix 5×5 size. Matrices differed not only in size, but also in the presence or absence of numerical estimates, the number of risk levels, or cell colors Table 3.2.

A typical "numerical" matrix with five risk levels is shown in Fig. 3.2. Similar matrices can be seen in Russian airlines.

To calculate the risk associated with the HF is necessary to determine the severity of the potential event that this HF may cause and the likelihood of this event. Such

Table 3.1 Quantitative risk assessment methods (in the wording of GOST R ISO/IEC 31010-2011)

№	Title	Method summary	Output	Use in GA
1	Human Factor Analysis (HRA)	Analysis of the impact of the human factor (HF) on the system and assessment of human errors affecting the operation of the system	The list of errors, causes, types, consequences, methods of reduction. Risk assessment	SMS of a/c of the Russian Federation, ASPPAP, MEDA (Boeing), THERP, ASEP, HFACS (NLR)
2	Toxicological risk assessment	Identification and analysis of dangers and possible ways of their distribution. Information on the level of harm to the environment and assessment of probability	The risk of exposure in the form of quantitative	There is no data
3	Fault Tree Analysis	The system failure (main event) is identified, and then the paths of its occurrence are depicted in the form of a logical tree diagram. Explore ways to reduce or eliminate the causes/sources of malfunction	Mixed or quality assessment	ASPPAP, Volga-Dnepr Airlines
4	Event Tree Analysis	To assess the probability of the implementation of events and their transition to other events, inductive conclusions are used	Scheme of occurrence of the final event	ASPPAP airline Volga-Dnepr
5	Analysis of the causes and consequences	Combines the methods of the fault tree and the event tree, taking into account the delay. The causes and consequences of the event may be investigated	– A set of minimum sections	ASPPAP airline Volga-Dnepr

(continued)

Table 3.1 (continued)

№	Title	Method summary	Output	Use in GA
6	Failure Types and Effects Analysis (FMEA) and Failure Types and Consequences Analysis (FMECA)	Identification of types and process of development of failure and its consequences. Several types of FMEA: project (or products) and their components, systems, process tech. service software. It may be accompanied by a critical analysis of the type of failure on a qualitative, quantitative, or mixed scale (FMECA)	– Assessment of the probability of the final event	ASPPAP, Volga-Dnepr Airlines
7	Maintenance aimed at ensuring reliability	A method for identifying and implementing a maintenance policy aimed at achieving the effectiveness and efficiency of the required safety, reliability, and cost-effectiveness of equipment	The solution of tasks in the field of those. service: monitoring tech. condition, scheduled repair and replacement, fault detection or current tech. service	Reliability systems and airline PLG
8	safety level analysis (LOPA)	The method allows us to evaluate controls and their effectiveness (analysis of barriers)	Recommendations on the use of controls and to reduce risk	ARMS (EASA), SMS Finair, SAS, Meridian
9	Bow tie analysis	A schematic way of describing and analyzing ways to implement risk (from danger to consequences and results), as well as analysis of management methods. The logic of the fault tree with (analysis of the causes) and the tree of events (consequences) are combined	Diagram of the main pathways of hazardous events and established prevention barriers with an assessment of the effectiveness	ARMS (EASA), SMS of airlines Finair, SAS, Meridian Air Shelf

(continued)

Table 3.1 (continued)

№	Title	Method summary	Output	Use in GA
10	Markov analysis	State analysis, used in the analysis of complex reconstructed systems that may be in various states, including those with poor performance	The probabilities of the system staying in various states, assessing the probabilities of failure and/or uptime	There is no data
11	Monte Carlo simulation	Used to establish changes in the system resulting from changes in input data, taking into account the distribution of input data and their relationship with the output data. Input data can be described as random variables. Use triangular or beta distributions	Characteristics, probability distribution or failure rate; state probabilities	"OverrunProbability" program, Volga-Dnepr airline
12	Bayesian analysis	The statistical procedure uses an a priori distribution of data to assess the likelihood of results. The accuracy of the results depends on the accuracy of the a priori distribution. Bayesian network models causation based on the analysis of probabilistic relationships of input data and results	– Values of output quantities	CATS (NLR, Holland), ASPPAP, Volga-Dnepr Airlines

Table 3.2 Probability categories of frequency during a raid of 100,000 h per year

Probability category	Frequency F	Frequency F for n = 100000
Category A	$F \leq \frac{1}{10N}.$	$F \leq 10^{-6}.$
Category B	$\frac{1}{N} \geq F > \frac{1}{10N}$	$10^{-4} \geq F > 10^{-6}$
Category C	$\frac{10}{N} \geq F > \frac{1}{N}$	$10^{=3} \geq F > 10^{-4}$
Category D	$\frac{100}{N} \geq F > \frac{10}{N}$	$10^{-2} \geq F > 10^{-3}$
Category E	$F > \frac{100}{N}$	$F > 10^{-2}$

Sample Risk Assessment Matrix

Severity**	Probability*				
Catastrophic Loss of life Loss of aircraft	5	10	15	20	25
Critical Serious injury Substantial damage	4	8	12	16	20
Serious Aircraft damage Minor injury	3	6	9	12	15
Marginal Operational effect Loss of employee time	2	4	6	8	10
Negligible System disruption	1	2	3	4	5
	Improbable (10^{-9}) Not likely to ever occur	**Remote** (10^{-7}) Likely to occur once every twenty years	**Occasional** (10^{-5}) Likely to occur once a year	**Probable** (10^{-3}) Likely to occur twice a month	**Frequent** (1.0) Likely to occur once a week

Fig. 3.2 Typical "numerical" risk matrix (from [19])

a path can be realized only by formulating, for example, by brainstorming a "list of federal orders". An example is given by us in [16, Sect. 4.2].

On the other hand, the identification of HF is associated with an analysis of the airline's operational activities. The main methods are given in [16]. This raises the question of how to reconcile the collected data and the proposed method of risk assessment.

Consider the capabilities of the risk matrix. There is a tendency not to indicate numerical values, but to designate cells with a number and a letter (Fig. 3.3).

The risk ranking recommended by ICAO is shown in Fig. 3.4.

It is acceptable to perform a quantitative interpretation of the matrix based on the following considerations [43]:

- for almost unbelievable events P = 0;
- for reliable events P = 1;
- for events with negligible damage S = 0,
- for events with catastrophic damage S = 1.

Risk to flight safety	Risk severity				
Probability	Disastrous A	Tolerable B	Significant C	Insignificant D	Negligible E
Often 5	5A	5B	5C	5D	5E
Sometimes 4	4A	4B	4C	4D	4E
Very rarely 3	3A	3B	3C	3D	3E
Unlikely 2	2A	2B	2C	2D	2E
Extremely unlikely 1	1A	1B	1C	1D	1E

Fig. 3.3 Risk matrix from ICAO SMM, 4th ed. 2018 year

Range of flight safety risk indices	Risk characterization	Recommended action
5A, 5B, 5C, 4A, 4B, 3A	Unacceptable	Immediately take measures to reduce the risk or stop the activity. Take priority actions to reduce safety risks to ensure that additional or improved prevention mechanisms are in place to reduce the safety risk index to an acceptable level.
5D, 5E, 4C, 4D, 4E, 3B, 3C, 3D, 2A, 2B, 2C, 1A	PERMISSIBLE	May be omitted based on safety risk reduction measures. The concept of this risk may require a management decision.
3E, 2D, 2E, 1B, 1C, 1D, 1E	ACCEPTABLE	Acceptable in its current form. No further measures are required to reduce the safety risk.

Fig. 3.4 ICAO SMM Risk Ranking Table 4-ed 2018

It should be noted that the range from 0 to 1 and must be divided into 5 equal parts. Then the matrix is shown in Fig. 3.2 will take the form of Fig. 3.5.

Unfilled cells correspond to an acceptable risk—"green risk", darker cells—to a risk acceptable provided that it is minimized to "yellow" risk, and the darkest cells—to an unacceptable "red" risk.

We can see that the entered numerical values of R do not always 3.5 correspond to the color of the cell. For example, cell 3B allows a value of R = 0.48, and the risk is "yellow" and the risk R = 0.38 in cell 4B is considered "red".

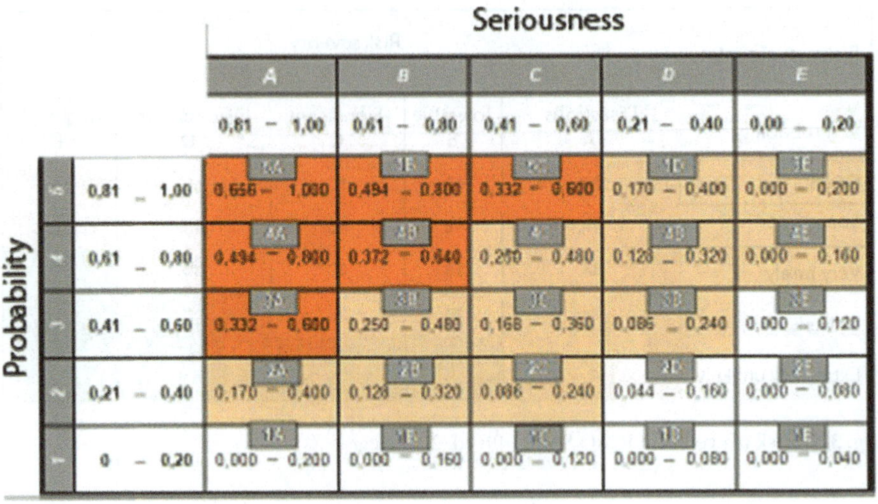

Fig. 3.5 Quantitative interpretation of the ICAO risk matrix

The ratio between the "green" and "yellow" risks is ambiguous, hence the different number of yellow and green cells in various documents. Therefore, in the USA FAA circular [20], there are two more green cells (due to the smaller number of yellow ones) than in the ICAO matrix.

In the manuals of aviation administrations and operators, matrices of a different dimension are found, as well as matrices with more than three risk levels ("intermediate" colors), matrices with curved borders of risk zones (the experience of Aeroflot-Cargo is given in [44]) and matrices based on three risk component in a Canadian document [17].

The paper [45] mathematically investigates matrices ranging in size from 2 × 2 to 5 × 5, used in various fields of technology, transport, and military affairs. We introduce the requirement of "weak consistency", which a matrix should have, formulated briefly as follows: "red" risk (in its quantitative interpretation) should always be higher than "green".

Two important theorems are proved.

1. To meet weak coherence, the "red" and "green" cells should not have common borders.
2. In the rightmost column and in the bottom row of the matrix, there should not be cells with the maximum risk level ("red" cells).

It is shown that the risk matrix does not always facilitate the adoption of better decisions with a random decision, notably when the P and S indicators are negatively correlated and allow one to reliably distinguish only "red" and "green" risks and only if weak consistency is satisfied.

An analysis of risk matrices is the focus of an article by a team of authors led by Makhutov [46], in which it was noted that matrices are certainly useful only with a positive correlation between P and S, which takes place, for example, in evaluating terrorist acts. With a negative correlation, estimates and decisions made using matrices are often erroneous. The aviation events are characterized by just a negative correlation between the probability and severity of the consequences, as was shown in [43], based on data on 5054 catastrophes in world civil aviation for 1952–2012 from the aviation insurance database.

In [45, 46], it was convincingly shown that with a sharply negative correlation between probability and severity, the choice of a solution based on a risk assessment by a matrix has no advantages compared to a random choice of a primary risk and the probability of error increases to 50%.

Does the ICAO methodology allow us to assess **the risk associated with an event**? Consider an example. The risk of the event is assessed: after type A, the aircraft stopped at the last runway plate after landing due to the reduced efficiency of the wheel braking system due to an error during maintenance. The crew used all means of braking. The result is the destruction of several aircraft. To assess the risk according to the "classic" option, we use the matrix: "seriousness/probability". A simple, at first glance, task upon more in-depth examination reveals an insurmountable shortcoming of this conceptual model.

The severity of the consequences of what event do we determine?

- the actual result (consequences)—a few torn tires?
- the most probable scenario of the accident: rolling out, accident?
- the most unfavorable outcome (according to the ICAO SMM 2009)—collision with an obstacle, destruction of aircraft, loss of life—a disaster?

The probability of what event we want to calculate and take into account:

- the same error during the technical inspection?
- a similar event—"almost rolling out"?
- real rolling out?

The answers to these questions depend on what events we should take from the database to calculate the frequency, and accordingly, the probability.

We can agree with *J. Nisula* that, considering an event (deviation), it is important to understand two things:

- the only factual part is the event in the form in which it happened and its real outcome, it is a "historical fact";
- "historical fact" in itself is not very interesting for risk assessment.

Next, we are interested in the ability of the HF, manifested in the event, to harm us in the future. Such an assessment will be based on subjective assumptions, and the assessment becomes uncertain. So, in our example, should we consider only Type A or also other types of airline aircraft? Should we consider other airfields? What events (deviations) can be considered "the same"? These questions are not always answered adequately, which makes the object of risk assessment unclear, and the whole process is erroneous.

Let us consider how the risk assessment system of one of the significant airlines described in [47] answers these questions.

A number indicates severity from 1 to 5 (event class). It is indicated that "events with the most irreversible consequences" belong to class 5. It turns out that the actual damage determines the class of the event. Then, for the example considered, the class of the event will be small, because the irreversible consequences—rupture of two or three tires—are small, although the potential consequences of the event are an accident.

If we approach the assessment of severity from these positions, then serious incidents related, for example, to a critical approach of aircraft in the air, will not have a class higher than 1 (fortunately, there are no irreversible consequences!).

Probability assessment (category) is related to the frequency of "the same" events for the "reporting period"—the number of flight hours per year.

Consider how this matrix can be applied in practice. For a sizeable Russian airline, in order to simplify the calculation of flight hours for a year, we can take N = 100000. Then the table for calculating the probability index given in [47] can be supplemented by a column of quantitative values (Table 2.3).

In order to get the category of an event, in any case, you need to calculate the frequency (we denote it in this section by F) according to the formula

$$F = \frac{N}{n}, \tag{3.1}$$

where: N is the number of evaluated events.

n is the flying hours of these N events.

Two assessment options are possible:

(1) by the number of events and the raid for the reporting period;
(2) according to the flying hours between the last and penultimate events.

To calculate according to option (1), select the reporting period. For practical reasons, the size of the reporting period can be either a month $n_m = 10000$ or a quarter $n_q = 40000$. For simplicity, we can take the raid for the reporting month, and for the quarter (raid may be uneven).

If one event occurs and the reporting period is one month, $F_m = 10^{-4}$, if the period is equal to a quarter, then $F_q = 0.25 \cdot 10^{-4}$. In both cases, the probability category is B. If the reporting period is one quarter, then category B will correspond to the number of events from 1 to 4.

Therefore, Category A will never be used.

If the period is a month and two to nine identical events occur, then $F_m = 2 \cdot 10^{-4} \div 9 \cdot 10^{-4}$ and probability category-C. The same situation if the quarter period and occurs from 5 to 39 events.

What events are supposed to be processed by such a matrix?

If these are incidents, then during the reporting period of one month, the overwhelming majority of incidents, regardless of the background, will automatically have the probability category "B", since it is infrequent for two incidents to occur in the airline for a month for the same reason. In the reporting period per quarter, all incidents will have the probability category "B", since there are no known cases when 5 or more "identical" incidents occurred in one airline within one quarter.

It turns out that the assessment of the probability category of an incident for the reporting period, suitable for risk management, loses its meaning.

If we are talking about deviations in general, then their number, of course, can be different. First of all, we note that in this case, Category A will never be used. However, questions will inevitably arise when establishing a class (seriousness).

Deviations of the same parameter (like incidents for the same reason) can have different severity (severity of consequences). For example, we have several landings with overloads on landing from 1.5 to 1.8 ("yellow level" according to Airbus guidelines [48]) and several—from 1.8–2.0 ("red level"). In calculating the frequencies, we can consider them "the same", but what class of event will be assigned to this set of deviations?

When calculating according to option (2), the frequency is calculated based on the flying hours n_I between the event in question and a similar event in the past

$$F = \frac{1}{n_I}. \tag{3.2}$$

This approach involves conducting a risk assessment of each event separately. In this case, theoretically, all categories can be involved (from A to C). For incidents, this approach is acceptable, since each is handled separately, but it is not suitable for minor events and deviations. The total number of such deviations can be measured in tens and hundreds per month, and they need to be processed together. Here we need to use computer data processing. If we assume that a computer program will be created that will assess the risk of each deviation according to option 2, then in order to make decisions, in any case, it will be necessary to calculate the "total risk" according to the type of events for a specified period.

Let this period be one month, and in one of the months, two deviations are noted with an interval of one day. If the flight hours between them are n1 = 300 h, then according to formula (3.2), we have F = 0.0033, which corresponds to category D. If in the next month similar two deviations occurred with an interval of 25 days n1 = 7500 h, then F = 0, 000133, which corresponds to category E. From this simple example, it can be seen that the method does not allow an objective comparison of the risk of deviations.

Further, as expected to analyze the totality of small deviations. If we try to add together the values of their risk, then the probability of addition will be taken into account twice and the severity only once.

Here is another example of the use of the matrix for the quantitative assessment of risk according to the results of activities proposed by professors of the MSTU CA Zubkov and Prozorov [49].

They propose an extended classification of events compared to PRAPI-98, which can be used in risk management systems. Eight classes of events were distinguished—from "precursors of low-risk incidents"—"PI-V" to catastrophes "K". Each class is assigned a degree of danger r_i inversely proportional to the permissible probability of these events, calculated based on the airworthiness standards of the AP-25 [50]. Accordingly, the degree of danger varies from 10^{-5} to 1. At the same time, the authors transformed the reasonable probabilities of special situations of AP-25 into the probabilities of events most relevant to these situations and added four classes of events for which there are no corresponding situations in AP-25.

The concept of "normalized event recurrence" Q_i is introduced, which is equal to the permissible probability of events of different classes from 10^{-2} for PI-B to 10^{-7} for K. Based on monitoring the airline's activity, it is proposed to calculate the risk indicator for the period U as

$$R_{\Sigma} = \frac{\sum_U r_i n_i}{T} \tag{3.3}$$

where r_i *and* n_i hazard indicator and the number of events of the i-th class; T-flying hours for the period U.

Further, it is not specified how the obtained values will be used in the matrix, which is proposed for assessing acceptability, the results of calculating the risk R_{Σ} by formula (3.3) are not presented.

Rows of 8 × 8 matrix constitute event classes, and the columns are the values of "normalized event recurrence". Therefore, the estimates "Acceptable", "Analysis", "Unacceptable" of the 3-color matrix refer to the events of each class, depending on their repeatability over the control period, which is calculated as

$$q_i = \frac{n_i}{T}$$

Thus, the method can be used to assess the level of S in the airline for the period. How to use the method for assessing the total risks R calculated by the formula (3.3) is not explained.

It can be concluded that the methodology of the ICAO does not fundamentally allow us to perform the risk monitoring of past events.

Another drawback of the approach, ICAO is that not considered barriers or protection of the well-known scheme of Professor Rizon [51], which is a vital concept in the prevention of ACCID.

How to evaluate risk reduction with the impact of barriers, but which can be set? For this step, no guidelines and evaluation will be very subjective.

The cause of the problems that occur when attempting a quantitative risk assessment using the matrix of ICAO is that this matrix is originally not intended for any quantitative calculations. That any of the risk matrix is qualitative instruments explicitly stated in [52] and several other works by well-known authors, e.g., in [53, 54]. Recent years increasingly used the matrix that emphasizes the qualitative nature of the assessments: for example, a matrix of airline "Aviashelf" given by us in [16].

Attempting to quantify risk using the matrix of ICAO based on statistics has arisen, in our view, for two reasons:

1. A natural desire to measure the risk of a literal interpretation of the definition of safety. As shown above, the ICAO standard is that external (government) assessment of the level of S in the airline (also risk assessment) should be based on established and agreed with the aviation authorities with objective statistical and qualitative indicators. **The calculated indices (indices) of risk used in the airlines cannot be a criterion for comparative evaluation of the level of risk.** They should be seen only as a performance intended solely for internal use in the airlines.

2. Incorrect understanding of the purpose of the frequency range indicated in the probability indicators in some risk management guides. These ranges are provided only to assist the expert in choosing an indicator and are not intended for any statistical calculations. At the airline level, when processing statistics of incidents and deviations, we do not deal with frequencies of less than 10^{-5}, so indicators related to "almost unbelievable events" and "improbable events" remain unclaimed.

The ICAO Risk Matrix may be used solely for expert judgment of future events. That is why ICAO does not set any standards for risk matrices.

The ICAO SMM 2009, Annex 1–3 to Chap. 5 contained examples of its use only for predicting the risks of alleged hazards. Unfortunately, there are no such examples in the new SMM [3].

Evaluations of past events (deviations) should be carried out using other technologies using unique methods, as noted, for example, in the work of the ECAST expert group [55].

Additional difficulties in understanding are created by the fact that the ICAO SMM [3] gives two different versions of the matrices, and there are numerous inaccuracies in the translation of the SMM into Russian.

3.2.4 *Review of Triple-Component Models for Safety Risk Management*

As noted, one of the drawbacks of risk models, ICAO is that it does not deal explicitly with barriers preventing events. In the practice of the airline is the redundancy of aircraft systems and crew actions for the prevention of the development of the script of ACCID. In the model, ICAO assumes that it should be considered when choosing the indicators of probability and severity. However, some of the guides and techniques prefer to talk about a three-component risk model.

As examples of GOST R 51814.2-2001, one of the guidelines on risk management of transport Canada [17], the practice of risk management in the airline "Aeroflot-cargo" [44], a methodology for occupational risk assessment in [56], studies on considering the vulnerability of complex technical systems in [57].

The program "INDICATE" the Australian Transport Safety Bureau (ATSB) offers the identification process HF taking into account barriers [23]:

- identify the potential danger of the airline, which can threaten the safety of passengers;
- assess the seriousness of the hazards;
- identify existing constraints;
- evaluate the effectiveness of each remedy; and,
- identify additional constraints.

In [16] describes known among the specialists of the "risk calculator", which is a simple computer program for calculating risk—asking directly on the computer screen levels of probability, impact, and consequences of the derived risk profiles. We can see that this model is not a two but three-dimensional matrix.

The third component is in different sources, documents, and programs are called in different ways: assessing the impact, parry, susceptibility, survivability, vulnerability, etc.

Consider approaches to assessing risk in two of these documents.

GOST R 51814.2-2001 describes a method aimed at the prevention of defects in technical objects and reduces the negative consequences of them by foreseeing failures and their analysis on the stages of design, development, and operation of structures and processes.

This approach is implemented in the following sequence:

- make a list of all potential types of defects of a technical object or process to take into account the experience of manufacturing and testing of similar objects, and experience and human error in the operation, maintenance, and repair;
- determine the possible adverse effects from each potential defect conduct a qualitative analysis of the severity of the consequences and quantify their significance;
- determine the causes of each potential defect and assess the frequency of occurrence of each cause by anticipated operating conditions, maintenance, and repair;
- sufficiency envisaged in the technological cycle of operations aimed at the prevention of defects, quantitatively assess the possibility of preventing the defect in the detection of the reasons at the stages of production and operation of the facility;
- quantitatively assess the criticality of each defect (with its reason), the priority number risk (PNR), PNR, and high are finalizing the design and production process, as well as the requirements and rules of operation to reduce the criticality of this defect.

Risk priority number RP in the adopted in this book, the notation is expressed as

$$RP^* = P^* S^* O^*, \tag{3.4}$$

where

P^*—defect probability indicator: quantitative assessment of the share of the product (of its total output) with a defect of this type;

S^*—the significance of the potential defect: a qualitative or quantitative assessment of the alleged damage from the defect;

O^*—defect skip probability indicator: quantitative assessment of the share of products with a potential defect of a given type, for which the provided monitoring and diagnostic methods will not allow to identify this potential defect or its cause if they occur.

The formula (3.4) multiplies the points (ranks) of the significance of the criteria set by experts. Points (for P*, S*, O* values from 1 to 10 are suggested) can be set according to prepared tables or scales, where the correspondence of the values of points and characteristics of the criteria is established.

Each type of defect can have several causes and several consequences. Accordingly, several PNRs are determined, and all the consequences are described. A critical boundary must be set in advance for PNR.

The Canadian manual [17] offers a similar option, where the risk calculation is written as follows:

$$R^* = P^* S^* V^*, \tag{3.5}$$

where: R^*—an indicator of the likelihood that a certain sequence of events will occur that entail specific consequences;

S^*—an indicator of the severity of the consequences;

V^*—vulnerability indicator (in [17] it is called exposure), which shows how vulnerable the system is to a given sequence of events.

Besides the fact that three components are taken into account, the method differs from the others in that the indicators P^*, S^*, and V^* are assigned the values 0, 1, 2, 3, and 4. Thus, the minimum value of at least one of them nullifies the risk. Otherwise, the ranking of P*, S* roughly corresponds to the ICAO approach.

When $R^* = 0 \div 1$ (minimal risk)—you can continue the activity after discussing the elements of risk.

When $R^* = 11 \div 30$ (moderate risk)—the continuation of activities after taking action to reduce the risk is allowed.

When $R^* > 30$ (high-risk)—is considered a threat to safety, activity ceases. Activities do not resume until measures are taken to reduce risk to an acceptable level.

The analysis of the applicability of risk assessment and risk management methods for S showed that:

- in the practice of S risk management in aviation enterprises, a small part of more than 30 existing methods is used, which is explained by the specifics of aviation activity;
- foreign developments (CATS, etc.) are not widely used in Russian airlines due to the limited data they process and the technical difficulties of implementation;

- the ICAO matrix remains the main tool for assessing risks, the use of which is limited to qualitative assessments; the matrix does not solve the problem of quantitative risk assessment according to operational activity data and does not directly take into account the vulnerability of the system, i.e., existing safety measures or safety barriers.

Therefore, this book proposes an approach to risk assessment based on a synthesis of existing industry and general scientific developments, taking into account the experience of several airlines involved in various types of aviation activities.

3.3 Estimation of the Flight Risk in Terms of Technocratic Conception

3.3.1 Model of Damage Distribution for Different Flight Scenarios

From the perspective of a general approach to the analysis of options for the development of hazardous events at high-risk objects, given in [57], within the framework of a probabilistic model, a flight can be represented as an implementation of one of the possible scenarios (Fig. 3.6).

Absolutely successful flight (zero damage) is the process of transition of the air transport system from the initial state D to the final state I_O (we are not considering

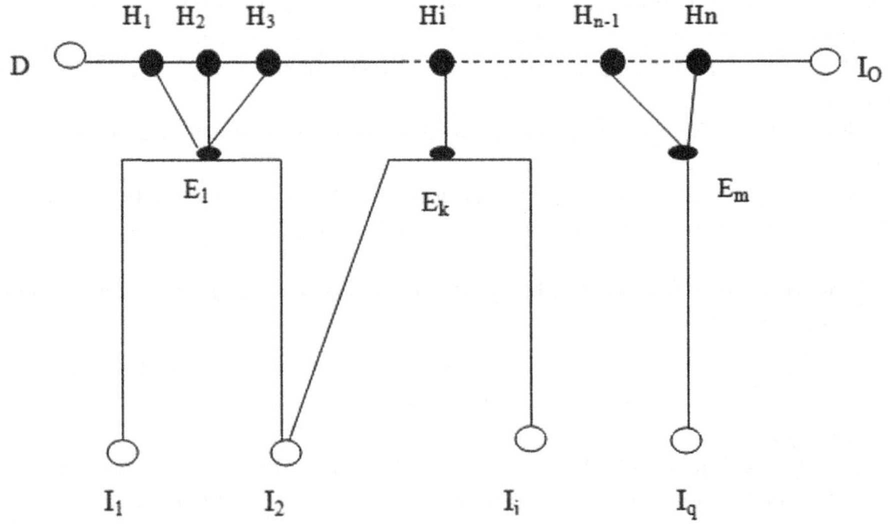

Fig. 3.6 The use case diagram of the flight

intermediate events E yet). Under the influence of HFs designated as H, the scenario deviates from the calculated trajectory; as a result, any of q possible events (flight outcomes) may occur I_j.

To simplify, let's imagine that each outcome I_j is described by only two parameters: the probability P_j of its occurrence and the amount of damage S_j in cost terms (for example, in US dollars), which will be inflicted on the airline. All flight outcomes form a whole group of events

$$\sum_{j=1}^{q} P(I_j) + P(I_0) = 1.$$

The product of the probability of each flight outcome with the damage can be considered as the risk associated with this outcome (event)

$$R_j = P_j \cdot S_j \tag{3.6}$$

The sum of the products (3.6) is the mathematical expectation of a random value of damage M (S), which is one of the possible characteristics of **the integral flight risk**

$$R = M(S) = \sum_{j=1}^{q} P_j S_j, \tag{3.7}$$

If we imagine an infinitely large number of flight outcomes, we come to a continuous random value of damage, for which the risk is like M (S)

$$R = M(S) = \int_{-\infty}^{+\infty} S f(S) dS,$$

where *f(S)*—probability density distribution of damage.

For practical tasks, not only the average damage values, but also the dispersion characteristics (dispersion or standard deviation) are essential. Figure 3.7 shows a possible form of two damage density distribution curves with the same average values of M (S), but with different variances.

The probability that the damage does not exceed any value is graphically represented by the area under the distribution density curve, limited to the right by this damage value. The gray area in Fig. 3.7 shows such a probability for the average value M (S) of the curve f2.

It is preferable to use either confidence intervals or quantiles of order P, where P is the probability (a quantile of the order of 0.95 is damage that will not be exceeded with a probability of 0.95). In Fig. 3.7, these quantiles are denoted as $S_{[0,95]}$. The figure shows that this characteristic reflects the nature of the distribution of damage.

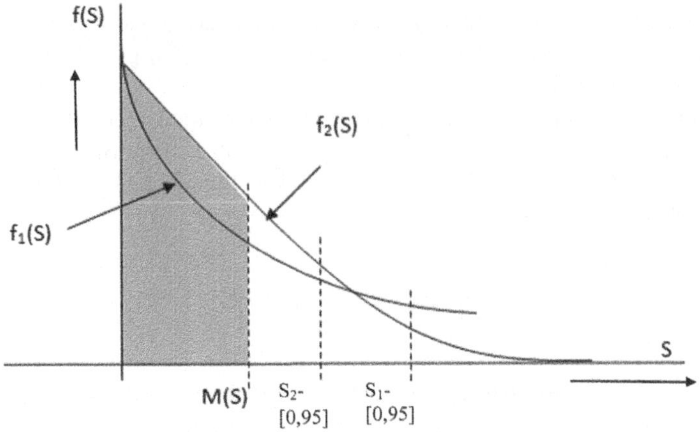

Fig. 3.7 Damage probability density functions

The practical application of this approach is challenging because:

- as noted above, parametric distributions are usually weakly connected with the real world;
- an obstacle to the use of nonparametric methods of calculation is a large degree of uncertainty of flight scenarios.

3.3.2 Vulnerability for Safety Concerns as an Important Aviation Enterprise Characteristic

Consider a diagram of flight scenarios in rice. 3.6 in more detail in terms of the development of each scenario.

The experience of investigating aviation events, studying the results of general studies on the development of accidents at hazardous facilities [9, 11, 57, 58] and foreign developments [59, 60] shows that the chain leading to airspace usually includes the following elements:

- the impact of HF (Hi), which is the initiating event;
- intermediate event Ek;
- escalation of the process leading to the final event Ij.

For example, a bird entering the engine in the climb is the HF (initiating event H_i), which leads to engine failure—(intermediate event E_k); as a result, an emergency landing on the departure aerodrome (end event I_j) with a specific damage S_j is performed.

The development of scenarios for each HF is determined not only by the intensity of its impact but also by the ability of the system to withstand HF. In military affairs, this characteristic is called "survivability" [31], and in [61], the characteristic "system stability" is used.

For risk management tasks, it is more convenient to use the opposite characteristic of the system: the degree of safety from exposure to the HF or the degree of exposure to the impact of the HF.

By analogy with the general analysis of complex, high-risk systems, we introduce the concept of "**Vulnerability of an Airline Enterprise for Hazard Factors**", because vulnerability is usually understood as openness, the susceptibility of the system to various negative influences.

The vulnerability of an airline (designation V) for a hazard factor is the conditional probability of the final state of the airline IJ going beyond the specified area of permissible states of the state Ω_m space in the case of exposure to the HF

$$V = P[(\|I_J - I_0\|) > \varepsilon_0 | H]. \tag{3.8}$$

The general method of adaptive adherence to restrictions on the components of the state vector of a dynamic system was developed at the MSTU GA [62]. The equations of the multidimensional state vector of the system in the general form are obtained, and the task is set of automating constraints with the help of a "predetermined, relay-triggered control". The approach can be used in assessing vulnerability, but so far, it has not been possible to fill the proposed model with specific content adequately.

The most significant difficulties are associated with obtaining an analytical, functional dependence for the parameters and finding their derivatives when considering complex constraint surfaces.

Therefore, at this stage, we use a simplified approach, in which the vulnerability is determined by the presence or absence of protection or "safety barriers". Barriers are naturally divided into two groups: "prevention barriers" and "parry barriers" (see Fig. 3.8).

Terms have the following meanings.

Hazard factor (HF)—a condition, object or activity that could lead to injuries to personnel, damage to equipment, structures, disruption of the interaction of system elements, loss of material assets or reduced ability to perform the prescribed functions.

Initiating event (IE) is the result of a hazard.

Safety barrier—a means (measure) of safety of two types.

Prevention barrier—a measure that prevents an intermediate event (InE).

A parry barrier—a measure that impedes the development of a substation to the level of an accident (ACCID) or an event with great damage.

Промежуточное событие—ПС *An intermediate event—InE*—development stage, when the ACCID can be avoided if the barrier to parry is available and involved.

In accordance with the adopted pattern of the development of the event, the vulnerability of the ATS is determined by two characteristics:

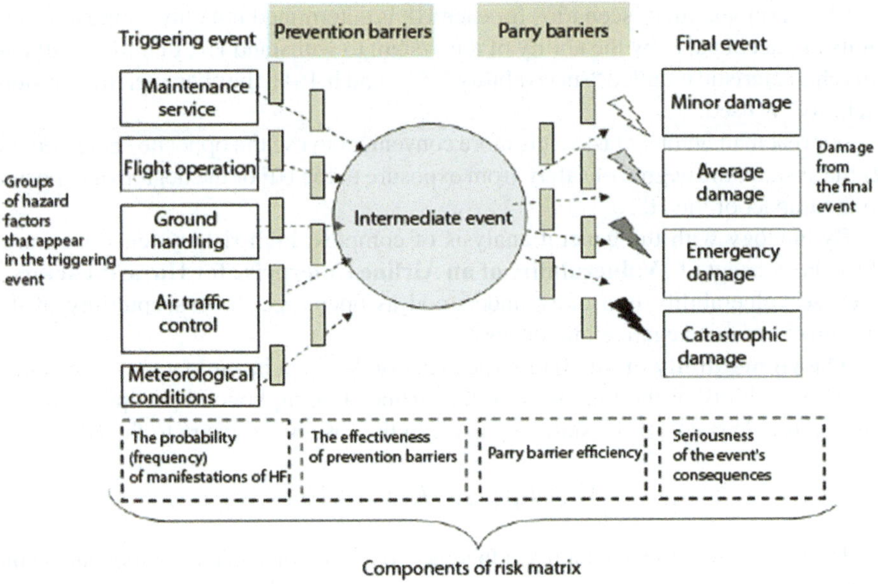

Fig. 3.8 Aviation event development scheme

1. the conditional probability of the occurrence of an intermediate event E in the case of exposure to hazard factor H;
2. the conditional probability of the occurrence of the final event I, provided that the intermediate event E.

It is vital that the same HF (H_i) can lead to different intermediate events (E_k), which, in turn, lead to different final events (I_j). The probability of any final event (I_j)

$$P(I_j) = \sum_{i=1}^{n}\sum_{x=1}^{m} P(H_i) \cdot P[E_K|H_i] \cdot P[I_j|E_K] \tag{3.9}$$

In matrix form, formula (3.9) has the form

$$P(I_j) = [P(I_j|E_1); P(I_j|E_2); \ldots P(I_j|E_m)]^T \times \begin{bmatrix} P[E_1|H_1] & P[E_1|H_2] & \ldots & P[E_1|H_n] \\ P[E_2|H_1] & P[E_2|H_2] & \ldots & P[E_2|H_n] \\ \ldots & \ldots & \ldots & \ldots \\ P[E_m|H_1] & P[E_m|H_2] & \ldots & P[E_m|H_n] \end{bmatrix} \times \begin{bmatrix} P[H_1] \\ P[H] \\ \ldots \\ P\{H_n\} \end{bmatrix}$$

The probability vector of all non-calculated flight outcomes

$$\begin{bmatrix} P(I_1) \\ P(I_2) \\ \dots \\ P(I_q) \end{bmatrix} = \underbrace{\begin{bmatrix} P[E_1|H_1] & P[E_1|H_2] & \dots & P[E_1|H_n] \\ P[E_2|H_1] & P[E_2|H_2] & \dots & P[E_2|H_n] \\ \dots & \dots & \dots & \dots \\ P[E_m|H_1] & P[E_m|H_2] & \dots & P[E_m|H_n] \end{bmatrix}}_{[V_A]\text{ - matrix of prevention barriers}} \times \underbrace{\begin{bmatrix} P(I_1|E_1) & P(I_1|E_2) & \dots & P(I_1|E_m) \\ P(I_2|E_2) & P(I_2|E_2) & \dots & P(I_2|E_m) \\ \dots & \dots & \dots & \dots \\ P(I_q|E_1) & P(I_q|E_2) & \dots & P(I_q|E_m) \end{bmatrix}}_{[V_R]\text{ - parry barrier matrix}} \times \begin{bmatrix} P(H_1) \\ P(H_2) \\ \dots \\ P(H_m) \end{bmatrix}$$

We can see that the transition to final states is described by two matrices, which can be interpreted as two components of the vulnerability as follows:

$[V_A]$—matrix of "**prevention barriers**", whose components are conditional probabilities of occurrence of InE under the influence of HF

$$[V_A] = \begin{bmatrix} P[E_1|H_1] & P[E_1|H_2] & \dots & P[E_1|H_n] \\ P[E_2|H_1] & P[E_2|H_2] & \dots & P[E_2|H_n] \\ \dots & \dots & \dots & \dots \\ P[E_m|H_1] & P[E_m|H_2] & \dots & P[E_m|H_n] \end{bmatrix}.$$

$[V_R]$—a matrix of "**parry barriers**" whose components are conditional probabilities of reaching individual final states I_j when intermediate events E_k

$$[V_R] = \begin{bmatrix} P(I_1|E_1) & P(I_1|E_2) & \dots & P(I_1|E_m) \\ P(I_2|E_1) & P(I_2|E_2) & \dots & P(I_2|E_m) \\ \dots & \dots & \dots & \dots \\ P(I_q|E_1) & P(I_q|E_2) & \dots & P(I_q|E_m) \end{bmatrix}.$$

$[V_A]$ matrix is the transition matrix from the vector HF $\{H\}$ to the vector of intermediate events $\{E\}$

$$\begin{Bmatrix} P(E_1) \\ P(E_2) \\ \dots \\ P(E_m) \end{Bmatrix} = \begin{bmatrix} P(E_1/H_1) & P(E_1/H_2) & \dots & P(E_1/H_n) \\ P(E_2/H_1) & P(E_2/H_2) & \dots & P(E_2/H_n) \\ \dots & & \dots & \dots \\ P(E_m/H_1) & P(E_m/H_2) & \dots & P(E_m/H_n) \end{bmatrix} \times \begin{Bmatrix} P(H_1) \\ P(H_2) \\ \dots \\ P(H_n) \end{Bmatrix}. \tag{3.10}$$

Since there are much fewer intermediate events than HF ($m << n$), the use of formula (3.10) allows us to reduce the dimension of the problem of calculating the probabilities of non-calculated flight outcomes (final events).

The matrix $[V_R]$ is the transition matrix from the vector of intermediate events $\{E\}$ to the vector of "final events" $\{I\}$

$$\begin{Bmatrix} P(I_1) \\ P(I_2) \\ \dots \\ P(I_q) \end{Bmatrix} = \begin{bmatrix} P(I_1/E_1) & P(I_1/E_2) & \dots & P(I_1/E_m) \\ P(I_2/E_1) & P(I_2/E_2 & \dots & P(I_2/E_m) \\ \dots & \dots & \dots & \dots \\ P(I_q/E_1\} & P(I_q/E_2) & \dots & P(I_q/E_m) \end{bmatrix}$$

The event development record in a short form has the following format:

$$\{I\} = [V_A] \cdot [V_R] \cdot \{H\},$$

or

$$\{I\} = [V] \cdot \{H\},$$

where $[V] = [V_A] \cdot [V_R]$—general vulnerability matrix for this model.

If we have a vulnerability matrix $[V]$, we can write risk value as

$$R = \{H\} \cdot [V] \cdot \{S\} \tag{3.11}$$

where $\{S\} = \{S[I_1]; S[I_2]; \dots S[I_n]\}$—damage vector (severity of consequences), the components of which are the damage values of the final states.

The product of the HF vector and the vulnerability matrix gives the vector of final states $\{P[I]\} = \{P[I_1]; P[I_2]; \dots; P[I_n]\}$, whose components are the probabilities of the system reaching various final states?

Then we get the usual risk formula (as a generalization of formula (3.1)) in the form of the product of two factors, one of which characterizes the probability of the onset of final states, and the second—damage (severity of the consequences of the event)

$$R = \{P[I]\} \cdot \{S\}$$

Vulnerability is the most crucial characteristic of any high-risk system when building risk management systems.

Following [57], it is recommended that work on reducing vulnerability be begun with an analysis of the construction of a matrix of parry barriers $[V_R]$; this allows one to identify critical intermediate events.

Next, we consider a matrix $[V_A]$ that allows us to determine the most dangerous HF.

Since the construction of accurate mathematical models for all flight scenarios at this stage of the development of science is not possible, **this book proposes the implementation of this "classical" approach** by constructing detailed scenarios for some of the most critical events in the form of "trees of development of aviation events."

Along with the "classical" probabilistic approach, **a simplified approach** is proposed that allows the use of **conditional indicators expressed by real numbers.** This approach is designed either to assess the risk associated with a particular financial statement (the group of financial statements) or to assess the risk for a specific type (aspect) of the airline's activities.

In this case, the following notation is used in this work:

R*—a conditional measure of risk;

P^*_{HF}—conditional indicator of the possibility of exposure to HF;

V*—a conditional indicator of vulnerability of an airline ($V^* \leq 0.1$);

S*—a conditional indicator of the severity of consequences (damage) when the final event occurs.

In this case, formula (3.1) takes the form

$$R^* = P^*_{FD} V^* S^* \quad (3.12)$$

In some cases, further simplification is applied, and a conditional indicator of the possibility of the final event is introduced $P^* = P^*_{FD} V^*$.

Then the conditional risk indicator is calculated as

$$R^* = P^* S^* \quad (3.13)$$

3.4 Structure and Principles of Safety Risk Management System Design in the Aviation Enterprise

3.4.1 Mathematical Description of Safety Risk Management Process

Attempts to describe the management of S and the risks of S using the achievements of the theory of control of complex systems are quite numerous [40, 47, 63]. However, for practical use in an airline, these results are not applicable.

The goal of this book was not to solve the problem of constructing a complete mathematical model of S risk management. To develop a methodology as a set of practically applicable methods is enough to use the general scheme and mathematical model in a simplified form as a description of the processes shown in the diagram of Fig. 3.9.

The system is based on the use of data on the activities of the airline and information from the authorized bodies of the CA and other external sources.

All information enters the information collection subsystem, which creates a structure with the selection of informative parameters and transmits information in a converted form to the control subsystem.

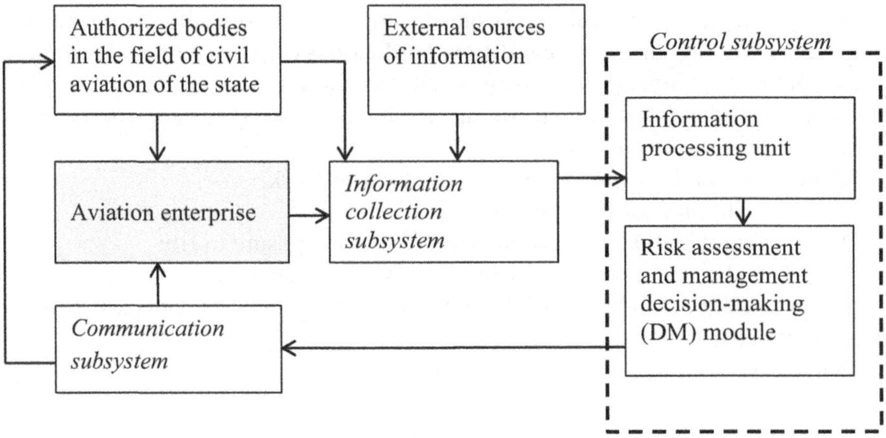

Fig. 3.9 S risk management scheme at the aviation enterprise level

A simplified diagram of the process of collecting and primary processing of information is shown in Fig. 3.10.

We assume that only the following M_J information arrays comprise the information base of the information collection system:

- M_1—information about deviations in the work of the units of the airline;
- M_2—crew errors and aircraft malfunctions according to the decryption of objective control means—flight information (FI);
- M_3—information from authorized bodies in the field of civil aviation;
- M_4—external information related to the S of the airline.

In the general case, the system can process k arrays of information that enter the control subsystem.

The management subsystem contains two main blocks: the information-processing unit and the module for the formation of managerial decisions and has a complex structure (see Fig. 3.11).

In the information-processing unit, the formation of risk vectors $\mathbf{R_J}$ occurs by converting the information M_j of each array by the operators F_j.

For example, based on information on deviations in the work of all departments, a risk vector of deviations $\mathbf{R_1}$ is formed, the components of which may be the risks of each department or activity, but based only on information from the array M1, and accordingly, not reflecting the real situation with risks by departments or direction activities. Therefore, the vectors $\mathbf{R_j}$ can be called "informational".

Further risk processing is carried out in the risk assessment unit, where the set of "information" risk vectors $\mathbf{R_J}$ is transformed into the risk vector of S $\mathbf{R}$ by the F^R operator.

Depending on the method chosen, and the level of management, the risk for S R may be:

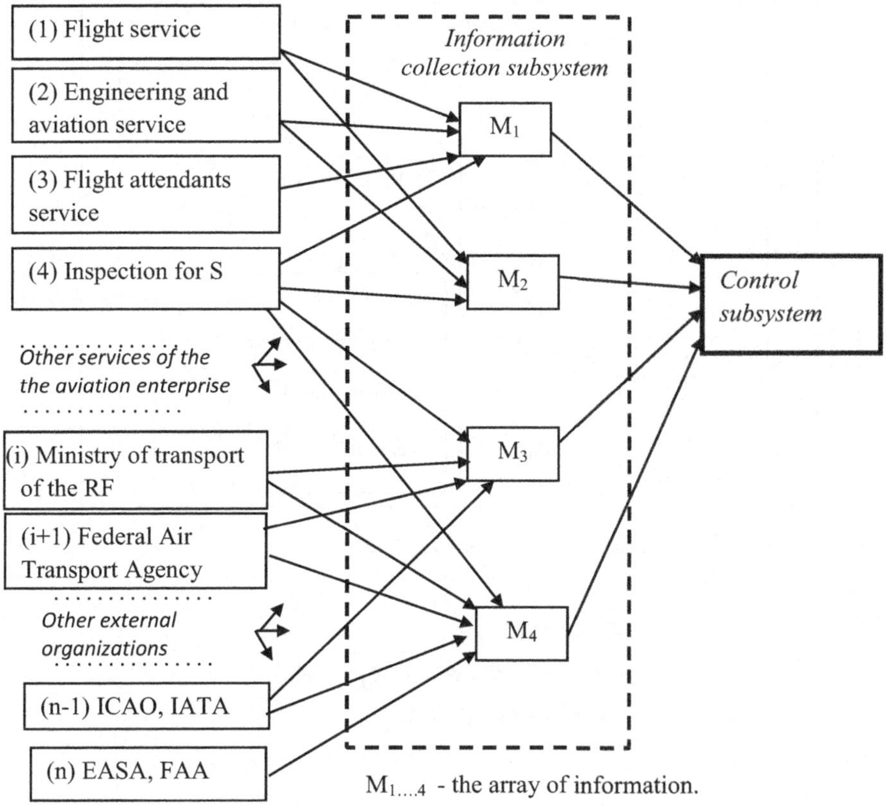

Fig. 3.10 Graph of the information collection and transmission system

- general risk of flight;
- the risk associated with a specific flight scenario (outcome);
- the risk of a particular type (direction) of activity of the airline;
- the risk associated with a particular financial statement or their combination.

The risk of S is directly used to develop management decisions (MD) and can be expressed as a characteristic of the probability distribution of damage, and a conditional indicator of risk.

The risk vector of S is transmitted to the input of the MD development block, where the vector of criteria for accepting risk in the areas of activity (types of events) **K** and the vector of typical MDs in the areas of activity **U** from the base of MD are transferred. According to these data, the matrix operator $\mathbf{F}^{\mathbf{Y}}$ carries out the development of a set (vector) of control decisions **Y**. The components of the vector **Y** are management decisions in certain areas (activities) of an airline: flight operation, aircraft maintenance, organization of transportation, etc.

Vector **Y** enters the communication subsystem, which closes the feedback circuit.

The generalized mathematical model of the risk management module has the form

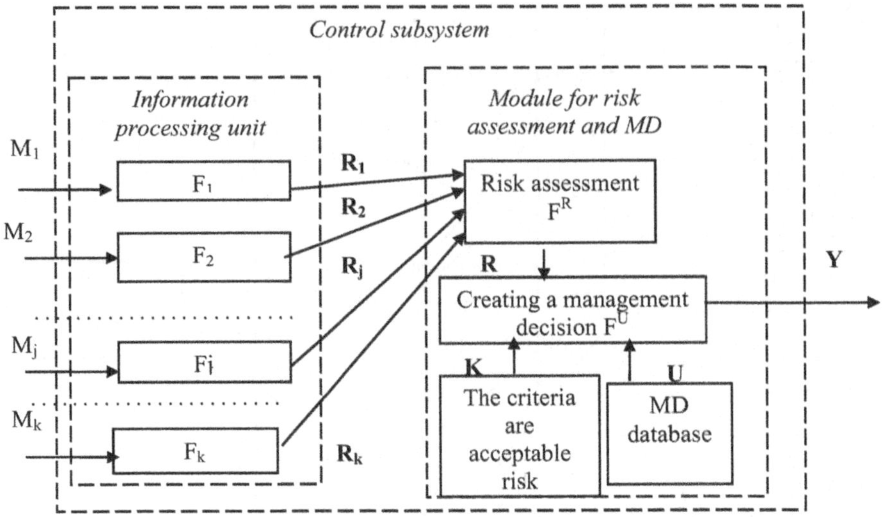

Fig. 3.11 Management decision vector formation scheme

$$\begin{aligned} \boldsymbol{R}_j &= F_j(\boldsymbol{M}_j); \\ \mathbf{R} &= F^R(\boldsymbol{R}_1, \boldsymbol{R}_2 \ldots \boldsymbol{R}_j \ldots \boldsymbol{R}_k); \\ Y &= F^Y(\boldsymbol{R}, \boldsymbol{K}, \boldsymbol{U}) \end{aligned} \tag{3.14}$$

3.4.2 Estimation of Safety Risk Acceptability on the Level of Aviation Enterprise

Following the proposed risk management model (3.14), in developing management decisions (MD), it is necessary to have criteria for assessing risk acceptability. As noted above, there is currently no state, and accordingly, industry standards for risk levels. This section sets out general provisions for developing risk tolerance criteria for an airline to measure the level of safety costs with the level of risk.

Considering S as part of the airline business, experts are faced with the problem of assigning this level. The task can be graphically represented (see Fig. 3.12) in the form of the well-known cost-safety ratio. The acceptable risk zone is formed in the minimum area of the curve "Total Costs", which is the sum of "Damage from an accident (ACCID)" and "Costs for ensuring safety (S)".

However, even in such a formulation, determining the acceptable risk zone seems to be a difficult task, since an indicator of the S level has not yet been established and, accordingly, it is impossible to obtain the original form of the curves of costs and damage as functions of the S level.

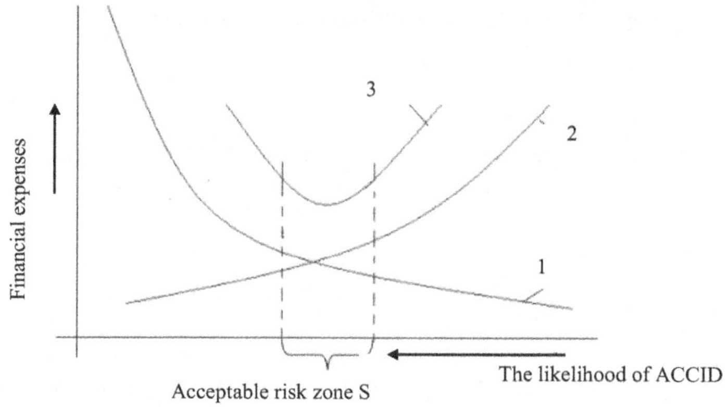

1- damage from an accident.
2- safety expenses.
3- total costs, as the amount of damage and expenses.

Fig. 3.12 Graphic representation of the acceptable risk zone for a S

In the practice of making decisions on a flight, the airline management necessarily applies the criterion of "cost-profit" in business activities [53, 64], which transforms into "cost-profit-risk" and is expressed by a simple formula

$$I - C - R > 0,$$

from which

$$R < I - C = P, \tag{3.15}$$

where R—risk as probable damage;

I—income from a new flight (charter program);

C—the cost of flight realization (program);

P—profit from the completed flight.

Formula (3.15) allows you to establish the minimum requirements for the level of acceptable risk: it should be less than the expected profit from the flight.

In the general case, when deciding on a flight, the leader assumes responsibility for the risks and/or transfers this responsibility to the insurance company. However, in practice, insurance never covers all the damage that a company may suffer in an aviation event. Therefore, when developing risk management systems, insurance is not considered a risk reduction measure and, as a rule, is not taken into account.

As in any economic activity, when carrying out air transportation, three financial results of making a decision on the flight are possible:

1. negative—with probability p we have the damage $\tau 1 = S$;
2. zero—the flight is not operated, risk aversion;
3. positive—with probability q we have the income $\tau 2 = I$.

The condition for the validity of the flight because of the flight

$$\tau = Iq - Sp > 0 \quad (3.16)$$

Since the economic results of a flight can be infinitely many, condition (3.16) is written as the mathematical expectation of the result of the flight

$$M(\tau) = \int_{-\infty}^{\infty} \tau f(\tau) d\tau > 0,$$

where f(τ)—probability density distribution of the expected financial result of the flight.

The occurrence of a risk situation is associated with the uncertainty of the future situation—the possibility of a different outcome of the flight. It is known that when making decisions, errors of the first and second kind may occur. In this case, a **mistake of the first kind** is a refusal to perform a profitable (safe) flight; **a mistake of the second kind** is the adoption of an unreasonable decision to perform an unprofitable (dangerous) flight. In both cases, the airline incurs losses.

The average loss of S_{av} due to an incorrect decision will be

$$S_{av} = \propto C + \beta(I - C),$$

where α and β—error probabilities of the 1st and second kind, respectively.

The probability of errors is determined by the capabilities of the risk management system for forecasting the situation, i.e., to reduce uncertainty. Assessing these probabilities is an essential task for any system based on a probabilistic risk model.

If inequality (3.15) is not satisfied, the airline takes measures to reduce the risk from level R to level R', spending the amount of funds C' on it and achieves a reduction of risk by $\Delta R = R - R'$, then the bright condition for the appropriateness of this measure is the inequality

$$\Delta R - C' > 0. \quad (3.17)$$

Inequality (3.17) itself can be applied when using probabilistic methods of risk assessment.

For methods based on conditional risk indicators, acceptable risk levels are established empirically.

3.4.3 Prognosticating of Management Decision Efficiency

The ultimate goal of applying the S risk management system is to develop management decisions (MD) that provide acceptable risk. We must have a methodology that predicts the effectiveness of the proposed MD and the issuance of recommendations to the head of the decision-making (HDM). There is an extensive literature on decision theory, various approaches are proposed; for example, in [65, 66].

Within the framework of the proposed methodology, a relatively simple method for predicting the effectiveness of MD based on the adaptation results for applying the experience of *the Commercial Aviation Safety Team* (USA) at the airline level has been developed.

The methodology is applicable both for risk in terms of value and for risk expressed by conditional indicators.

Without loss of generality, we consider the application of the methodology to assess the effectiveness of risk reduction measures in value terms for individual types of events.

Given the previously accepted notation, we introduce the following indices:

$i = \overline{1, n}$—indices of hazard factors (HF);

$j = \overline{1, m}$—indexes of types of aviation events (AE);

$k = \overline{1, u}$—indexes of management decisions (MD).

Let the risk of an aviation event (AE) of the j-th type R_j calculated according to formula (3.1) exceed the acceptable level.

A list of financial statements that made the most significant contribution to this risk is compiled or automatically generated.

Based on expert estimates [14], the contribution $G_{ij}\varepsilon(0, 1) \sum G_{ij} = 1, \sum G_{ij} = 1$, of each i-th HF to the risk of the j-th AE is calculated.

Several MD_k are proposed or formed, each of which is characterized by the following parameters:

– the effectiveness of the influence of the k-th MD_k on the i-th HF in the form of the probability that the i-th HF does not appear with the full implementation of the k-th MD Q_{ij}^{kP}; this probability can be estimated by statistics, by technical data of systems, or by expert means (for example, the effectiveness of installing on the ground early warning systems for ground proximity (GYWSGP) to compensate for deficiencies in the radio equipment of the aerodrome can be estimated from event statistics or by interviewing experts);

– the degree of implementation of each k-th MD_k designated as $M_k\varepsilon(0, 1)$, in the form of the share of crews trained, the part of the aircraft fleet where the units were inspected or new equipment was installed, etc.;

– the cost of selling the k-th MD in monetary terms C_k.

Based on the known formulas of probability theory [66], the following are calculated:

– part of the risk of the j-th AE attributable to each i-th HF

$$R_{ij} = R_j G_{ij};$$

– the probability that the i-th HF will not appear for the j-th AE when introducing one k-th MD

$$Q_{ij}^{k} = Q_{ij}^{kP} M_k;$$

– the likelihood that the i-th HF will not manifest itself in the implementation of all MD[1]

$$Q_{ij} = 1 - \prod_{k=1}^{u} \left(1 - Q_{ij}^{k}\right);$$

– the eliminated risk associated with the influence of the i-th HF for the j-th event after the introduction of one k-th MD

$$\Delta R_{ij}^{k} = R_{ij} Q_{ij}^{k};$$

– residual risk associated with the influence of the ith HF for the j-th event after the introduction of one k-th MD

$$R_{ij}^{k} = R_{ij} - \Delta R_{ij}^{k},$$

– the eliminated risk associated with the influence of the i-th HF for the j-th event after the implementation of all (several) MDs

$$\Delta R_{ij} = R_{ij} Q_{ij},$$

– the residual risk of the i-th HF for the j-th event after the implementation of all (several) MD

$$R_{ij}^{o} = R_{ij} - \Delta R_{ij},$$

– the eliminated risk of the j-th AE after the introduction of one k-th MD

$$\Delta R_{j}^{k} = \sum_{i=1}^{n} \Delta R_{ij}^{k},$$

– the residual risk of the j-th AE after the introduction of one k-th MD

$$R_{ij}^{k} = R_{j} - \Delta R_{j}^{k},$$

– the eliminated risk of the j-th AE after the implementation of all (several) MD

$$\Delta R_{j} = \sum_{k=1}^{u} \Delta R_{j}^{k},$$

– the residual risk of the j-th AE after the implementation of all (several) MD

$$R_j^o = R_j - \Delta R_j,$$

– total eliminated flight risk (of the airline) after the implementation of the k-th MD

$$\Delta R^k = \sum_{j=1}^{m} \Delta R_j,$$

– the total residual risk of the flight (airline) after the implementation of the k-th MD:

$$R^{ko} = R - \Delta R^k.$$

– the total eliminated risk of the flight (airline) after the implementation of all (several) MD

$$\Delta R = \sum_{k=1}^{u} \Delta R^k,$$

– the total residual risk of the flight (airline) after the implementation of all (several) MD

$$R^o = R - \Delta R,$$

– the share of the total risk of the flight (airline) eliminated by the k-th MD

$$\Delta R^k(\%) = \frac{\Delta R_{.k}}{R} 100\%,$$

– forecast of the overall effectiveness of the k-th MD

$$\Delta R_C^k = \frac{\Delta R^k}{C_k},$$

– the effectiveness of the k-th MD to reduce the risk of the j-th AE

$$\Delta R_{jC}^k = \frac{\Delta R_j^k}{C_k}.$$

The head of decision-making (HDM) is provided with an assessment of each SD and their combinations according to several criteria: reduction of the overall risk level, reduction of the risk of certain types, risk reduction per unit of funds invested in the MD, etc.

For practical application of the technique in the airline, a program was developed at MSExcel. An example of using the program to assess the effectiveness of real MD in an airline is given in App Annex location 1.

3.4.4 General Principles of Safety Risk Management System Design in the Aviation Enterprise According to the Proposed Methodology

The methodology provides for the development of various methods for managing the risk of S, subject to the following basic principles:

1. Isolation of three risk management loops in an airline: operational, tactical, and strategic, corresponding in general to the three hazard identification methods recommended by ICAO.
2. The separation of risk management methods into two types: methods for responding to events (determining the risk at the time of their occurrence), and prognostic methods of risk management.
3. The maximum possible use of real data on the operational activities of the airline, their structuring, and formalization to ensure the possibility of using relatively simple software tools for processing and analysis.
4. An expanded interpretation of the quantitative presentation of risk, allowing the use of conditional indicators.
5. Consideration of the properties of the system to counteract hazards through vulnerability assessment (assessment of the effectiveness of safety barriers).

An analysis of the airline's work shows that in the practical activities of S management (safety risk management), we can **select three control lines**.

We will call them "Operational", "Tactical", and "Strategic".

These circuits, to a certain extent, correspond to the three methods for identifying hazard factors recommended by the ICAO SMM [3], which are called "reactive", "proactive", and "predictive" in the Russian translation. The general control scheme is shown in Fig. 3.13.

Operational management involves a "risk assessment" associated with specific events or deviations. In essence, these are operational decisions on responding to dangerous events, which are immediately taken by responsible managers (PIC—Pilot In Command, shift supervisors, etc.).

Actions depend on the class of the event. At present, the classifier Rules for the investigation of accidents and incidents 98 [66], containing requirements for the investigation of aviation events, are in force. However, for the methodological support of decision-making, a classifier proposed by the specialists of the MSTU CA [49], which provides for 8 classes, can be adopted.

The implementation of the contingency plan also applies to this case (element 1.4 of the conceptual framework of the ICAO SMS [54]).

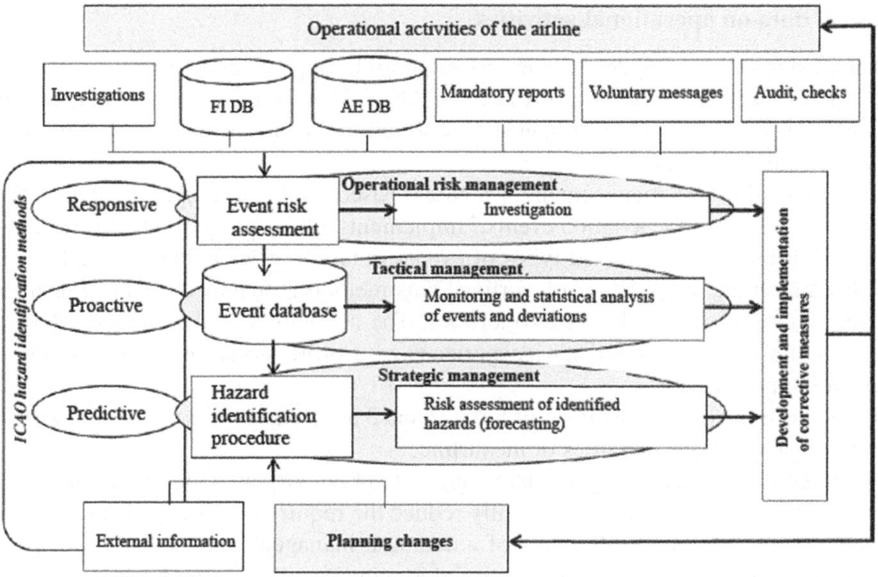

Fig. 3.13 Three lines of risk management at the enterprise

It is crucial here that the information about the event and the rejection should be entered into the database for use in other management methods.

An expert using special database-based programs with the provision of analysis results and recommendations for corrective actions implements ***tactical management*** through an ongoing risk assessment. Such an assessment can be performed weekly or monthly. Airlines, to one degree or another, have implemented precisely this part of the scheme, although they call this work an "event risk assessment".

Strategic management consists of conducting periodic (once a quarter, half a year, or a year) evaluations of the results of statistical analysis, data from other sources, including external ones, taking into account changes in activities, and making appropriate strategic decisions. It is for this task that various matrices of probabilities and consequences are adapted, including the ICAO matrix.

Separation of risk assessment procedures is associated with past events and risk prediction.

To speak of a "risk event" is possible only conditionally. Attempts to directly "fit" the event in the matrix ICAO does not give a positive result.

The methodology involves the development of different methods for risk assessment that existed when the event occurred and methods of forecasting as a risk assessment of an uncertain future. Of course, information about the events is the basis for forecasting.

Use of data on operational activities

The methodology provides for the fastest possible implementation of the principle of the SMM ICAO "effective SMS driven data." This data includes the data that is named in [66] "control and diagnostic information." If use of software products of different difficulty levels.

A method of risk management of S No. 1, based on the development of the "tree of development of the aviation events," implements a fully probabilistic approach to assessing the risk of selected types of events in terms of value based on all available data on the activities of the airline. This method is intended for large airlines operating high-risk, such as cargo airlines. The adaptation of the method requires considerable resources and long experience as a professional team of specialists, including programmers.

Methods of risk management of S No. 2 and 3 are based on the use of conditional risk measures on the activities of the airline.

Method 2, as method 1, uses large amounts of activity data, but the principle of conditional indicators can significantly reduce the required resources. However, the method involves the development of a database management system, so for a big company with professional programmers is required. The method can be used in large and medium-sized airline operating passenger flights.

Method 3 applies conditional risk measures with a clear separation of risk assessment procedures "past events" and risk prediction of hazards and a clear account of the safety barriers. The method can be used in medium and small airlines.

References

1. Pospelov GS (1980) Systems analysis and artificial intelligence. G. S. Pospelov. Moscow: Computing center of the USSR, 47 p
2. Pospelov DA (1986) Situational management: theory and practice. D. A. Pospelov. –M.: Science-CH. ed. Fiz.-Mat. lit., 288 p
3. A guide to safety management (SMM) (2013) Doc.9859. ICAO, 3rd edn, 300 p
4. Leveson NG (2011) Engineering a safer world. The MIT Press, Massachusetts, Cambridge, London, England
5. Nisula J (2009) Operational risk assessment. Next Generation Methodology, [Electronic resource] Mode of access: easa.europa.eu"essi/documents/ARMS.pdf
6. GOST R ISO/IEC 31010 risk Management-2011 (2011) Methods of risk assessment. Publishing House of Standards, 71 p
7. GOST R 51901.1-2002 risk Management (2002) Risk analysis of technological systems. Publishing House of Standards, 40 p
8. Safety of Russia (2008) Legal socio-economic and scientific-technical aspects. Risk analysis and safety management. In: (Methodical recommendations) team NA, Makhutov N, Pulikovsky KB, Shoigu SK (ed). Moscow: MGF "Knowledge", 672 p
9. Alexander LN et al (2001) Statistical methods analysis of the safety of complex technical systems: Textbook, Logos, 231 p
10. Kravets VA (1980) The Method of "fault tree" in the analysis of safety systems to the oil and gas industry. Internet registry, 40 p

11. Shvyrev V (1992) Probabilistic analysis of the safety of nuclear stations. Methods of execution. Yu. Shviryaev and others. –M.: IAE them. Kurchatov IV, 266 C
12. Beshelev SD (1980) Mathematical-statistical methods of expert assessments: textbook. Beshelev SD, Gurevich FG –M.: Statistics, 263 p
13. Drozdov AV, Spesivtsev AV (1994) Formalization of expert information in the logical-linguistic estimation of complex systems. Tech Cybern 2:89–96
14. Orlov AI Organizational and economic modeling: a tutorial: 3 h. Orlov AI. –M.: Izd-vo MGTU im. N. E Bauman, 2011, Part 2; Expert evaluation, 486 p
15. Barkalov SA et al (2008) Model of conflict management and risk management: monograph. Barkalov SA [and others] under the editorship of D. A. Novikov. –Voronezh: Scientific book, 495 p
16. Zubkov BV (2010) Theory and practice of risk management in aviation enterprises in the development of system safety management. Zubkov BV, Sharov VD. –M.: MGTU CA, 196 p
17. Risk Management & Decision Making in Civil Aviation, TP-13905B (2004) Transport Canada
18. System Safety Management Program (2004) FAA
19. Background and Fundamentals of the Safety Management System (SMS) for Aviation Operations (2006) ALPA, Second Edition, 2006. 20 Introduction to Safety Management System for Air Operators AC No 120-92, FAA
20. SMS Tool Kit Publication of the US Joint Helicopter Safety Implementation Team. September 2009, p 37 Safety, 2011. [Electronic resource] Access mode: www.IHST.org
21. Safety Management System Assessment Guide (2005) TP 14326E, Transport Canada
22. Proactively Monitoring Airline Performance INDICATE Australian Transport Safety Bureau [Electronic resource] Access mode: http://www.atsb.gov.au/publications/1999/sir199906_002.aspx
23. Materials of the seminar-meeting "SMS in the Civil Aviation Administration and Preparation for ICAO Verification by USOAP" (2008). [Electronic resource]. M.: FAVT, 1 electron.opt. Disk (CD-ROM)
24. Guidelines for providing information support for the automated safety system of the civil aircraft of the Russian Federation (ASOBP) (2002) –M.: LLC "Air Navigation Consulting Agency", 192 p
25. The IOSA Standards Manual (ISM) (2013) 7th edn, IATA
26. Lin P-H (2011) Safety Management and Risk Modelling in Aviation: Next Generation Infrastructures Foundation P. O. Box 5015 2600 GA Delft, The Netherlands
27. Lushkin AM (2010) Methodical provision of monitoring procedures in the system of safety management. abstract. dis.... Cand. technical Sciences: 05.22.14/ Lushkin Alexander. –M., 22 p
28. Matveev GN (2010) The method of proactive safety management in aviation enterprises.abstract. dis.....candidate of technical Sciences: 05.22.14/ Matveev Georgiy Nikolaevich. –M., 28 p
29. Ryabinin IA (2000) Reliability and safety of structurally complex systems. Ryabinin IA. –SPb.: Polytechnic, 248 p
30. Mozhaev AS (1999) Computer-aided logical-probabilistic modeling of technical systems (user Guide PC ASM version 5.0). Mozhaev AS, Alekseev AO, Gromov VN. –SPb.: Military engineering-technical University, 63 p
31. Sharov VD (2007) Synthesis of the functions of the a priori estimation of the level of safety of the upcoming flight for the group of factors of Environment. V. D. Sharov// Problems of safety. –No. 11, pp 13–23
32. Guziy AG (2007) Theory and practice of quantitative evaluation of the risk of an aircraft accident. In: Guziy AG, Grudzinsky VA (eds) Problems of safety, vol 9, pp 10–18
33. Itskevich AA (2009) Probabilistic-statistical model of the operation of aircraft: textbook. Itskevich AA, Kabakov PK. –M.: MSTU CA, 120 p
34. Probabilistic Risk Assessment Procedures Guide for NASA Managers and Practitioners (2002). NASA, Washington, DC 20546
35. Causal model for Air Transport Safety (2009). Final report. NLR, Amsterdam
36. Fedyukin VK (2005) Managing quality of processes. Fedyukin VK. –SPb.: Peter, 202 p

37. Logvin A.I (2010) Reliability and technical diagnostics. Part II "Technical diagnostics REO": a tutorial. In: Logvin AI, Yakovlev DA (eds) MSTUCA, 56 p
38. Evdokimov VG (2013) Monitoring and safety operations, taking into account changes in the functional properties and risk factors of complex technical systems (aviation systems): author. dis…. Dr. of Technical Sciences: 05.22.14. Evdokimov VG. –M., 38 p
39. CAP 712 Safety Management Systems for Commercial Air Transport Operations (2002). CAA UK
40. Unlocking the Potential of an SMS. Safety Digest, November-December 2005
41. Sharov VD, Balloons VD (2009) Some mathematical and logical restrictions on the use of risk matrices in the system safety management. Sci Bull MSTUCA 149:179–181
42. Kruzhalov AI (2008) The system of safety management based on risk management. Sci Bull MSTUCA 135:71–178
43. Cox L (2008) What is wrong with risk matrices? Risk Anal 28(2):497–512
44. Makhutov NA, Reznikov DO, Petrov VP, Kukseva VI (2012) The use of risk matrices during the risk assessment and prioritization of protective measures. Probl Safety Emer 1:82–92
45. Zubkov BV (2009) Safety management. Part I. Safety Program. Safety Management Guide: A Training Manual. In: Zubkov BV, Polyakov PM, Karmizov MV (eds) –M.: MSTU CA, 132 p
46. Flight Profile Manual, LOMS, Version 2.63, Airbus (2002)
47. Zubkov BV, Prozorov SE (2011) Methodology for risk assessment in the safety management system. Sci Bull MSTU GA 174:7–11
48. Aviation Rules (2004) Part 25. Airworthiness standards for transport category aircraft, 2nd ed, IAC, AVIAIZDAT OJSC, 240 p
49. Reason JT (1997) Managing the risks of organizational accidents. AP Company, Brookfield Vermont USA
50. Safety of Russia (2006) Legal socio-economic and scientific-technical aspects. Risk and safety analysis. In 4 parts. Part 1. Fundamentals of safety analysis and regulation. Scientific leadership Frolov KV. –M.: MGF "Knowledge", 640 p
51. Vishnyakov YD (2008) A General theory of risks: proc. A manual for students. ouch. proc. institutions. Vishnyakov YD, Radaev NN –2-e Izd., M.: Publishing center "Akademiya", 368 p
52. Selyukov VK (2008) Risk management organization: textbook.manual. Selyukov KV –M.: Izd-vo MGTU of N. E. Bauman, 188 p
53. Guidance on Hazard Identification ECAST (2009) [Electronic resource] Mode of access. http://www.easa.europa.eu/
54. Levashov SP (2009) The methodology of expert assessment of occupational risk. Life Safety 1:14–16
55. Makhutov N (2009) 57, N. And. Ensuring the safety of critical facilities based on the vulnerabilities. In: And N, Makhutov N, Petrov VP, Reznikov DO (eds) Saferty problems and emergencies. –M.: VINITI, №2, pp 50–69
56. Makhutov N (2006) 58, N. A. Development and application of methods of risk management in problems of technogenic safety and technical regulation. In: And N, Makhutov N (eds) Problems of safety and emergencies, № 1 –M: VINITI, pp 35–49
57. CAP 760 Guidance on the Conduct of Hazard Identification. Risk Assessment and the prediction of Safety Cases for aerodrome Operators and Air Traffic Controllers. CAA UK (2006)
58. Safety Management Systems for Flight Operations and Aircraft Maintenance Organizations—A guide to implementation, TP 13881, Transport Canada (2002)
59. Kulba VV (2011) The Use of scenario and indicator approach for managing survivability, durability, and safety of complex technical systems. Scientific edition. Kulba VV, Kononov DA, etc.– M. ICS RAS, 116 p
60. Vorobyov VV (2012) The method of adaptive aging restrictions on the components of the state vector of a dynamic system. Sci Bull MSTUCA 179:114–122
61. Sakach RV (1989) Safety. In: Sacac RV, Zubkov BV, Davidenko MF (eds) –M.: Transport, 239 p
62. Stupakov VS (2005) Risk-Management: manual. In: Stupakov VS, Tokarenko GS (eds) –M.: Finances and statistics, 288 p
63. Orlov AI (2006) Theory of decision-making. In: Orlov AI (eds) –Moscow: Exam, 573 p

64. Rykov AN (2009) Converged management and decision support. In: Rykov AN (ed) Moscow: IKAR, 245 p
65. Wentzel ES (1998) Probability theory: proc. for universities. –5-e Izd. ster. Wentzel ES. –M.: Vyssh. wk., 576
66. Rules of investigation of aviation incidents and incidents with civil aircraft in the Russian Federation (1998). PRAPI 98. Approved. The RF Government decree of 18 June 1998, No. 609. M: Aviaizdat, 140 p

Chapter 4
Definition and Monitoring of Safety Performance Indicators

4.1 Safety Performance Indicators as Criteria of Safety Risk Management Efficiency. Review of Safety Performance Indicators Applicable in Civil Aviation

The availability of objective quantitative indices of S and their established target and limit levels is a prerequisite for practical work of the State safety regulations (SSR) and the SMS service provider [1]. Therefore, the development of such indicators and methods of their monitoring is an integral part of the methodology of risk management of S. In this chapter, a review of S indices used in CA, and the main results of research on the development of an indicator of S for the SSR and of the operator's SMS.

ICAO approach, consisting of using multiple indicators of S, has long been used in domestic practice and theory for calculations of reliability and safety. For example, in [2], it is argued that the range of safety performance indicators should meet the following requirements:

(1) to be necessary for the safety features of the object, i.e., the indicators should be independent of each other.
(2) to be sufficient for safety features, i.e., a set must meet a range of tasks.
(3) to answer the previously formulated definition of safety.

In reliability analysis and probabilistic safety analysis [2–4] are used indicators that are divided into universal and particular.

Universal indicators characterize the properties of an object without taking into account its specificity and can be applied to any type of object.

The specific indicators characterize the safety of the object, reflecting the specific aspects of its functioning.

Example of a universal indicator is the probability of failure $P(t)$ associated with a function of time to failure $F(t)$ and density $f(t)$ dependencies:

V. D. Sharov et al., *Probabilistic-Statistical Methods for Risk Assessment in Civil Aviation*, Springer Aerospace Technology, https://doi.org/10.1007/978-981-16-0092-0_4

$$P(t) = 1 - F(t),\ P(t) = 1 - \int_0^t f(t)dt.$$

The probability of failure-free operation is summarized by the indicator "probability of no ACCID" and is directly related to such particular indicators as "probability of ACCID" and "probability of SS".

Two crucial indicators need to be mentioned.

The availability factor K_t is the probability that the object will be in a healthy state at an arbitrary point in time.

The operational readiness coefficient is K_{OR} as the probability that the object will be in working condition at an arbitrary moment in time, and, starting from this moment, it will work faultlessly for a given time interval. The formula determines this coefficient

$$K_{OR} = K_t P(t).$$

An important indicator of reliability or safety is the *criticality of failure* C, which means a certain number from a predetermined range (dimensionless scale), reflecting the consequences of failure.

If the failure is essential only in the economic aspect, then its criticality characterizes the reliability of the object. If the failure is critical, then the value C characterizes the safety of the object.

Reliability or safety indicator—the specific *average number of undesirable events*—or the mathematical expectation M(N) of the number of events (failures, accidents) per unit operating time during time T: $n = \frac{M(N)}{T}$.

This indicator is an analog of the private indicator "average number of ACCID per 100,000 flying hours" and characterizes the safety of the fleet of similar facilities.

In the practice of S management in CA, mainly private indicators are used. Here is a reasonable list of indicators from GOST [5].

(1)–the number of ACCIDs per 100 thousand flight hours; (2)–the probability of absence of ACCID; (3)–the probability of ACCID; (4)–the probability of a particular situation (SS) during the flight; (5)–the probability of fixed assets for one hour of flight; (6)–ACCID severity coefficient—the ratio of the number of disasters to the number of ACCID; (7)–SS hazard coefficient—the probability that the SS will lead to ACCID; (8)–the average raid on one ACCID.

Indicators (1), (2), (3), (4), (6), (7), (8) are currently used in various systems and programs for evaluating and forecasting S.

Another criterion for the danger of a particular situation may be, according to [6], the probability of non-pairing by the crew or air defense systems of the consequences of S. In the calculation, normative probabilities, according to ACCID-25 [7] are used for special situations (SS): catastrophic (CS), emergency (E), complex (C), and flight conditions complications (FCC). The approach corresponds to [5].

Since the probability of occurrence of each SS is regulated, the probability of unpaired situations is determined from the assumption that each of them gives the same probability of a catastrophic situation Q

Table 4.1 Probability thresholds in special situations

Probability	Special situation			
	Catastrophic Situation (CS)	Emergency (E)	Difficult situation (DS)	The complication of flight conditions (CFC)
$q_1(q_2)$	$10^{-9}(10^{-7})$	$10^{-7}(10^{-6})$	$10^{-5}(10^{-4})$	(10^{-3})
$r_1(r_2)$	1	$10^{-2}(10^{-1})$	$10^{-4}(10^{-3})$	(10^{-4})

$$Q = q \cdot r = const,$$

where q—the probability of occurrence of SS.

The table of threshold values of r obtained based on this formula and standards [7] is given in Table 4.1.

Correspondence of the actual S level to acceptable is determined by comparing the calculated values of r and q with the standard ones. The method involves determining the probability of manifestation of each factor in flight, which in itself is a challenging task. It must be borne in mind that the existing standards [7] relate to the SS due to equipment failures. The application of this approach to the human factor and external influences is a separate task. The approach to evaluating S using the S probability indicator, proposed and developed in [6], has found practical implementation in SMS of several Russian airlines.

For example, Aeroflot uses the integrated S coefficient

$$S(\%) = \left(1 - \frac{n_{CFC} \cdot K_{CFC} + n_{DF} \cdot K_{DF} + n \cdot K_E + n_{CSE} \cdot K_{CS}}{N}\right) \cdot 100\% \tag{4.1}$$

where n_{CFC}, n_{DS}, n_E, n_{CS}—the absolute number, CFC, DS, E и CS;

K_{CFC}, K_{DS}, K_E, K_{CS}—coefficients that take into account the share of influence of each type of special situation in flight on the S as a whole (Table 4.2).

Q—the total probability of an SS occurring in flight;

P—probability of non-occurrence of SS in flight (%).

In this indicator, the ratio of the severity of events is taken equal to the ratio of the probabilities of the SS from [7], which requires a separate justification. In practice, airlines use a simplified approach: the forerunner is CFC, and the incident is DS, which eliminates a severe differentiation of events.

Table 4.2 Coefficients of influence of SS on S in the Aeroflot SMS

SS type	Q	P (%)	K
CFC	10^{-3}	99,9	1
DS	10^{-4}	99,99	10
E	10^{-6}	99,9999	1000
CS	10^{-7}	99,99999	10000

Table 4.3 The relative severity of the effects of ACCID

Person	Severity	Aircraft	Severity
Injury	1	Damage	1
Mutilation	4	Crash with recovery	6
Fatality	64	Loss	20

A more sophisticated system for calculating the safety factor operates at Aviashhelf Airlines [8], where two coefficients are adopted.

1. Flight operation safety factor—C_{f}/

$$C_f = \frac{O_f}{N},$$

where O_f—assessment of deviations from requirements for the period under review;

N—number of flights for the period under review.

Safety factor for technical operation—C_m

$$C_m = \frac{O_m}{G},$$

where O_m—summarized estimate of deviations from the established requirements for maintenance service (MS) for the period.

G—scope of maintenance work for the period under review.

The adopted system was analyzed in [9]. It can be noted that the severity ratio of various aircraft adopted by Aviashhelf Airlines is close to the airspace identification in [6] for class 1 airplanes, which were obtained by processing statistical data for the period 1984-1991. In [10], data are presented on the ratio of the severity of the consequences of ACCID (see Table 4.3).

Data were obtained on the percentage of losses from ACCID: loss of aircraft is considered as ½ of the cost of a new aircraft and makes up about 33% of the total loss structure, search and rescue operations −18%, loss of profit −34%, compensation to passengers −15%.

Doctor of Technical Sciences Rukhlinsky [11] proposes a new criterion for evaluating S in the article. The use of correction factors taking into account insurance payments for damage from ACCIDis proposed. It is indicated that the proposal for the implementation of the criterion was accepted for consideration at the 36th Assembly of ICAO.

In [12, p. 61] as a S indicator, it is proposed to introduce "the probability of a heavy commercial aircraft crash for 1 flight" and establish an acceptable level of $3 \cdot 10^{-7}$.

Summing up, we note that either currently the CA of the Russian Federation does not have an official S indicator at the state or at the corporate level; although in the official S analysis [13] the number of ACCIDs per 1 million departures is named as the "main global S indicator".

Table 4.4 shows the system of indicators from the annual S analyzes in the CA of the Russian Federation [13]. We can see that for different groups of aircraft and types

Table 4.4 Statistical S indicators for different types of CA, types of transportation and aircraft used in the FATA reports

	Absolute indicators						Relative indicators				
	ACCID	Dis	Average cat for 3 years	Dead			Per 100,000 H.		Deaths per 1 million transported	ACCID for 1 million flights	Dis per 100 thousand hours for 3 years
				Total	Onboard	On land	ACCID	Dis			
Commercial aircraft											
Certifications. according to FAP-11 (FAP-246), all-transportation	X	X	–	X	X	X	X	X	X	–	–
Regular	X	X	X	X	X	X	X	X	X	X	–
Irregular	X	X	–	X	X	X	X	X	–	–	–
Aerial work	X	X	–	X	X	X	–	–	–	–	–
Airplanes - all	X	X	X	X	–	–	–	–	–	–	X
Aircraft up to 10t	X	X	–	X	–	–	–	–	–	–	–
Aircraft over 10t	X	X	–	X	–	–	–	–	–	–	–
All helicopters	X	X	X	X	–	–	–	–	–	–	X
Helicopters up to 10t	X	X	–	X	–	–	–	–	–	–	–
Helicopters over 10t	X	X	–	X	–	–	–	–	–	–	–
Certificate by FAR AR-all	X	X	–	X	X	X	–	–	–	–	–

(continued)

Table 4.4 (continued)

	Absolute indicators						Relative indicators				
	ACCID	Dis	Average cat for 3 years	Dead			Per 100,000 H.		Deaths per 1 million transported	ACCID for 1 million flights	Dis per 100 thousand hours for 3 years
				Total	Onboard	On land	ACCID	Dis			
General aviation (GA)											
Certified - all	X	X	X	X	X	X	–	–	–	–	–
Aircraft	X	X	–	X	X	X	–	–	–	–	–
Helicopters	X	X	–	X	X	X	–	–	–	–	–
Autogyros	X	X	–	X	X	X	–	–	–	–	–
Not having certificate	X	X	–	X	X	X	–	–	–	–	–
Unregistered aircraft	X	X	X	X	X	X	–	–	–	–	–

of air transportation, different indicators are used. It will be shown below that the use of a large number of indicators is also characteristic of other aviation organizations.

4.2 Analysis of Problematic Issues in the Definition and Application of Safety Statistical Indicators

For a correct comparison of the level of S in Russia with other States and the global level of S according to ICAO, IATA, and other sources, it is necessary to use data on events with groups with similar characteristics and the same types of traffic. However, the reports of States and organizations use different classification entirely.

In Russia, by order of the Ministry of Transport of the Russian Federation dated 20.06.1994, No. DV-58 aircraft are divided into four classes depending on the maximum certified take-off mass: 4th grade—up to 10t aircraft, helicopters up to 2 m; 3rd grade—airplanes 10-30t, helicopters 2-5T; 2-class—planes 30-75T, helicopters 5-10t; first class-75T aircraft, helicopters more than 10t. Additionally, the VC of the Russian Federation introduced the definition of a light aircraft with a massless 5700 kg and ultralight—less than 495 kg. Note that in the reports of MAK under "light" refers to aircraft with a maximum take-off mass (MTOM) less than 10T., and it may be the difference for the statistics in Russia and the CIS.

ICAO and EASA have adopted other classification: up to 2250 kg, from 2250 to 3750; from 3750 to 5700 kg; from 5700 kg to 27.5 T, 27.5 T to 272 T; and more 272 T.

The IATA report provides data for the aircraft of Western production more than 15t, and for sun East production does not include the group aircraft take-off weight [14].

ICAO and EASA, as a rule, analyzes statistics for aircraft weighing more than 2250 kg. In the reports of ICAO type [15] provides data on ACCID if you are only regular flights.

Insurance companies use their classification. For example, firm Ascend results in the calculation of statistical parameters for commercial flights on the aircraft seat 15 or more.

NTSB USA leads the statistics for commercial aircraft number of seats more than 10 [16].

Firm Boeing tracks statistics for aircraft with MTOM > 6000 lbs.

Usually, the calculation of commercial air travel, which includes regular and non-scheduled flights, cargo, and passengers, but some organizations use different classifications. So IAA results in the calculations the relative amount of AE and disasters for the following categories:

- all aircraft of civil aviation, including the GA;
- all aircraft of civil aviation, in addition to GA;
- all heavy (Maximum Take-Off Mass (MTOM) > 10T) cargo planes;
- heavy (MTOM > 10T) passenger aircraft;
- aircraft (MTOM < 10T) without IAA;
- all helicopters to civil aviation without IAA.

ICAO Council's reports and in statistical yearbooks [15] provides data for all commercial aviation world and 6 regions (Africa, Asia, Europe, Latin America, Middle East, North America) with MTOM > 2250 kg using the following indicators:

1. The number of accidents per 100,000 departures by years and average for 6 years;
2. The number of deaths of passengers per year and average over 6 years.

European Aviation Safety Agency (EASA) uses the indicator number of accidents per 10 million departures. It is calculated for 7 regions: North America, South America, Africa, Europe (members of EASA), Europe (not members of EASA), Western and Central Asia, South and South-East Asia, East Asia, Australia, and New Zealand. This is a standard indicator for scheduled commercial flights for aircraft with an MTOM > 2250 kg. All disasters are divided into 28 categories. Looks separately at the indicators for helicopters and caller ID, as well as for light aircraft (MTOM < 2 250 kg.). Indicators are calculated separately for events associated with airports and air traffic (ATM). In this case, if the airports are only considered up, then the events of the ATM, along with an indicator of the absolute number of ACCID of different categories, are calculated and the indicators for incidents. All incidents are divided into 5 classes according to severity: A, B, C, D, E. the design is carried out relative indicators of the number of incidents in categories A and 1 million hours and 1 million departures. Uses simple moving average with a smoothing period of 3 and 5 years. The same principle is used in some countries [17].

Eurocontrol uses the following metrics [18]: the absolute number of ACCID associated with ATM, the relative number of incidents related to ATM categories A, and b (similar to EASA) per 1 million flight hours. The calculations are performed and the types of events. Accounted for aircraft with an MTOM of more than 2250 kg, but it is not specified whether the calculations only for commercial, or for all aircraft.

The International Air Transport Association (IATA) monitors the level of S and publishes the results annually in their reports, which leads to the following statistical parameters [14]:

1. The number of ACCID with the destruction to the aircraft of Western production to 1 million flights for all airlines and separate airlines-members IATA.
2. The number of ACCID global commercial aircraft 1 million flights.
3. The total number of disasters and the number of deaths separately for jet and turboprop aircraft of Western production for years.
4. The number of transported and the number of deaths per 1 million carried separately for jet and turboprop aircraft of Western production.
5. The damage from ACCID separately for jet and turboprop aircraft of Western production.

For each of **the 8 regions**: Africa (AFI) Asia Pacific (ASPAC), CIS (CIS), Europe (EUR), Latin America and the Caribbean (LATAM), Middle East and North Africa (MENA), North America (NAM) North Asia (NASIA), indicators are calculated:

1. The number of ACCID with the destruction for the jet aircraft of Western production to 1 million flights.
2. The total number of ACCID regions for all aircraft 1 million flights.
3. The number of ACCID member airlines of IATA and non IATA.

In the report [14] are identified 12 types of ACCID and 18 phases of flight.

For each type of ACCID is calculated the total number of ACCID, the percentage of these ACCID total numbers of each region, and each phase of flight. Note that the IATA approach to the calculation of the indicators does not allow to correctly comparing the levels of S, because of the following drawbacks:

1. The differences in the formation of the initial data for the calculations of indicators. Some indicators are calculated for all of the Western and Eastern production. By definition in [14] for the aircraft of Western production includes all commercial aircraft built in Western Europe, America, and Indonesia with MTOM $\geq$ 15000 kg for jet aircraft with MTOM $\geq$ 5700 kg for the turboprop.

To sun East production is commercial aircraft manufactured in the CIS countries and China without specifying MTOM, i.e., the calculation takes into account AE with all "Eastern" aircraft, and small aircraft of Western production is not taken into account.

2. In the report [14] on page 15 shows two figures with the data on accident rates by region for the aircraft of Western production and all of VS (Western and Eastern production). Indicators in the second image for different regions more than in the first figure 2–8 times, which should indicate the worst situation with sun East production. But upon consideration, it is evident that the first figure—the comparative figures for the ACCID with the loss of the sun, and the second—the performance of all ACCID. Obviously, to compare these levels is incorrect.

The member states of the European Civil Aviation Conference (ECAC) have introduced **the SAFA** (Safety Assessment of Foreign Aircraft) program—checks of foreign aircraft with regard to air safety based on the results of which the "Airline Risk Index" is calculated.

When checking, forms are used where comments are recorded, ranked in three categories by severity level [19].

Risk is calculated as

$$R = \frac{0,25N_{C1} + N_{C2} + N_{C3}}{I} \tag{4.2}$$

where N_{C1}, N_{C2}, N_{C3}—the number of comments of the first, second, and third categories;

I—the number of aircraft inspections for a given period.

Based on this technique, in the Russian Federation, by Order of the Federal Aviation Administration dated 11.09.09 No. 401 "On conducting inspections of civil aircraft at airports in the Russian Federation", an aircraft inspection procedure similar to SAFA and a risk calculation method according to formula (4.2) was introduced. The same order in the Russian Federation established "risk levels"—preventive 2.0

and maximum 2.5, upon reaching which sanctions can be applied to the airline. The ECAC does not establish specific levels, but draws up lists of airlines with high levels of risk and imposes restrictions.

The significance of this coefficient in airline risk management is sometimes exaggerated, and at low values, it is said that the risks for S in the airline are insignificant. Despite the importance of SAFA inspections, their results are only one of the sources of information in risk management for S.

A comparative analysis of indicators for different aircraft can be performed for specific types of events. In [20], data were obtained at one of the meetings that the relative number of rollings out of the runway when landing on aircraft of western production is half that of eastern production.

The results of the study on statistical indicators, classification of aircraft, and types of air transportation by different organizations are shown in Table 4.5.

4.3 Method of Definition and Monitoring of the State Safety Performance Indicators

To assess the level of S should at least have:

- the indicator (metrics);
- approved algorithm for the calculation of this indicator (indicators);
- set the value of the indicator as acceptable levels of S.

Statement of the concept of an acceptable level of S in the basic ICAO documents ICAO in the period since 2006 (since the release of the first edition of the SMM) at present has undergone specific changes. Before entering Annex 19 acted in Annex D of "Acceptable level of S "Annex 11" Air Traffic Services", which has been invoked in other Annexes, in particular in Annex 6 "Operation of air transport". This Supplement D contained detailed provisions which were by definition the ICAO Standards, i.e., had a higher status than the recommendations of ICAO SMM [21]. The former variant of the concept, ICAO has been analyzed in detail in [9].

Annex 19 [1] not concretize indicators for SSR and SMS, calculation methods, and target levels. All specific recommendations shifted to the SMM ICAO [21]. Stated that when developing indicators and target levels, the state should use the data of the state system of data collection and processing for S (*Safety data collection and processing system—SDCPS*). Besides, the target values of indicators must be installed and threshold levels. Thus, it is assumed the definition of some of the permissible intervals of values of S.

SMM ICAO recommends as indicators of the S in the SSR at the initial stage to use the relative number of ACCID and SI. It is proposed to install two types of indicators: serious consequences (High Consequence Indicators) and less severe consequences (Lower Consequence Indicators). These concepts have replaced the

Table 4.5 Aircraft categories, types of air transportation, and used statistical indicators

Organization	Aircraft category (*M* = *MTOM*)	Types of air transportation	Used S statistics			
			All ACCID	ACCID with loss of aircraft	Disasters	Registration of the dead
ICAO	M > 2.25t	Regular passenger	For 1 million flights	–	For 1 million flights	Amount
IATA	Western-made aircraft M > 15T	All	For 1 million flights	For 1 million flights	Amount	1 million transported.
	Eastern production	All		–	Amount	Amount
EASA	M > 2.25T	All commercial	Amount	–	Amount	–
		Reg. passenger	Amount		At 10 million. flights	
Boeing	Reactive aircraft west.p-va M > 60,000 lb	All commercial	For 1 million flights	For 1 million flights	For 1 million flights	Amount
FATA RF	All sun	Commercial	For 100 thousand hours	–	For 100 thousand hours	1 million transported.
		Regular	At 1 million departures	–	–	–
	All helicopters	All commercial	For 100 thousand hours		For 100 thousand hours	For 100 thousand transported.
IAA	All aircraft, including GA	All	For 100 thousand hours	–	For 100 thousand hours	Amount
	All aircraft without GA	All				
	All aircraft M > 10T	Commercial				
	Pas. ACFT M > 10T	Commercial				
	ACFT < 10t without GA	All				
	Helicopters without GA	All				

(continued)

Table 4.5 (continued)

Organization	Aircraft category *(M = MTOM)*	Types of air transportation	Used S statistics			
			All ACCID	ACCID with loss of aircraft	Disasters	Registration of the dead
NTSB USA	10 or more passengers (14 CRF 121)	All commercial	For 100 thousand hours, 1 million. miles, 100 thousand departures	For 1 million hours	For 100 thousand hours, 1 million. miles, 100 thousand departures	Amount
		Regular		–		
		Irregular				
	Less than 10 passengers (14 CRF 135),	Regular				
		Irregular				
	GA	All				

concepts of "level of safety" (Safety measurement) and "Measurement of safety performance" (Safety performance measurement) of the ICAO SMM, 2nd ed. 2009.

To calculate the indicator of severe consequences in [21] is recommended to use the relative number of the amount of ACCID and serious incidents (SI) per 1000 flight hours per month/quarter. The coefficient "number of events per 100,000 h" seems more convenient and the calculation formula takes the form

$$K_{St-i} = \frac{\sum (N_{ACCID} + N_{SI})}{n_i} \times 100000 \quad (4.3)$$

where: N_{ACCID}, N_{SI} the number of ACCCID and SI for a certain period;

n_i—total flying time of all state operators for the i-th period, hour.

As a target K_{S_t} for the next year, it is proposed to use the average monthly value for the last year, reduced by 5%, i.e., average monthly value of last year

$$K_{St-LY} = \frac{\sum_{i=1}^{12} K_{St-i}}{12} \times 100000,$$

target level for next year

$$K_{St-LY} = \frac{\sum_{i=1}^{12} K_{St-i}}{12} \times 100000., \quad (4.4)$$

To calculate threshold levels in the current year K_{St}^{A} in [21] is proposed to use the standard deviations (SD) for the previous year

$$\sigma = \sqrt{\frac{\sum_{i=1}^{12}\left(K_{St-h} - K_{St-LY}\right)^2}{11}}.$$

It is proposed to assign three alarm levels in the current year

$$K_{St}^{A1} = K_{St}^{T} + \sigma,\ K_{St}^{A2} = K_{St}^{T} + 2\sigma,\ K_{St}^{A3} = K_{St}^{T} + 3\sigma. \quad (4.5)$$

A state body should form a set of indicators. As the SSR develops, indicators for events with lesser consequences can be used and indicators used in quality and reliability systems. The indicators should be used for comparison by year, periodically, it is necessary to conduct an audit of the indicator system.

We give an example of calculating the indicator. Formulation of the problem

(1) using the recommendations of ICAO and data on the flight hours and events in 2011 calculated the target state level of S and threshold levels for the indicator of serious consequences for 2012;
(2) according to actual data in 2012, calculated monthly values of this indicator and maintain a monitoring schedule.

As initial data, we use official analyses of the Federal Air Transport Agency [13] for 2011 and 2012 (hereinafter—"Analysis 11" and "Analysis 12").

Flight hours n in commercial aviation of the Russian Federation for 2011 calculated based on Analysis 11 by the number of ACCIDs and the indicator of the number of ACCIDs per 100,000 h (0.69) as

$$n = \frac{18}{0,69} \cdot 100000 = 2608696.$$

Let us take the conditional distribution of this plaque with seasonal variations ± 10% of the average monthly flight of 2608696/12 = 217391 h.

The distribution of events (ACCID and SI) is based on the Annex to Analysis 11. The indicator is expressed in the number of events per 100,000 h (Table 4.6).

According to the calculation results, we have

- average value for 2011 $K_{St\text{-}LY} = 1{,}4$; SD in 2011 $\sigma = 1{,}04$.

According to (4.3), with a 5% increase in S, we have the target S level of the current year:

$$K_{St1}^{T} = 0{,}95 K_{St-LY} = 0{,}95 \cdot 1{,}46 = 1{,}39,$$

and by formulas (4.4), we calculate threshold levels:

Table 4.6 Calculation of the state monthly S indicator for 2011

	Month	n, hour	Event	Status	N	K_S
1	Jan	195652	01.01-Surgut, Tu-154	Dis	1	**0,51**
–	Feb	195652	–	–	0	**0,00**
2	Mar	195652	15.03-Mi-8T	Crash	4	**2,04**
3			25.03-Il-76	SI		
4			25.03-Tu-204	SI		
5			28.03-A-319	SI		
6	Apr	217391	16.04 Yak-40	SI	3	**1,38**
7			28.04 An-2	SI		
8			17.04 An-148	SI		
9	May	217391	11.05 An-26	SI	4	**1,84**
10			05/19 X-32 "Snipe"	Dis		
11			17.05 An-2	SI		
12			14.05 Yak-42	SI		
13	Jun	217391	20.06 -Tu-134	Crash	6	**2,76**
14			02.06-An-26	SI		
15			18.06 Mi-2	Crash		
16			24.06 An-2	SI		
17			05.06 A-330	SI		
18			11.06 Yak-42	SI		
19	Jul	239130	5.07-Mi-8	Crash	9	**3,76**
20			11.07-An-24RV	Dis		
21			9.07-SP-30	Crash		
22			12.07-Mi-8	Crash		
23			15.07-Mi-8	Crash		
24			14.07 Yak-40	SI		
25			25.07 An-24	SI		
26			25.07-Mi-8	Dis		
27			19.07-Mi-8	Dis		
28	Aug	239130	08.08-An-24	Crash	4	**1,67**
29			17.08 An-24	SI		
30			9.08-An-12	Dis		
31			29.08-Mi-8	Crash		
32	Sep	239130	09.09-Yak-42	Dis	2	**0,84**
33			23.09 X-32 "Snipe"	Crash		
34	Oct	217391	14.10 Tu-134	SI	2	**0,92**
35			27.10 An-24	SI		

(continued)

Table 4.6 (continued)

	Month	n, hour	Event	Status	N	K_S
36	Nov	217391	25.11 Mi-171	SI	2	**0,92**
37			28.11 Mi-8	SI		
38	Dec	217391	20.12-Mi-26	Dis	2	**0,92**
39			9.12 A-320	SI		
	Total	**2608692**			**39**	

$$K_{St}^{A1} = 1{,}39 + 1{,}04 = 2{,}43;\ K_{St}^{A2} = 1{,}39 + 2{,}08 = 3{,}46;\ K_{St1}^{A3} = 1{,}39 + 3{,}12 = 4{,}5.$$

Based on the data obtained, a monitoring screen for the S indicator (see Fig. 4.1) for 2012 is plotted, the target level and threshold levels are applied according to the "traffic light" principle (data for 2011 are shown with a dashed line).

During 2012, monthly K_{St} is calculated based on actual data according to formula (4.3), i.e., Analysis 12 is in progress. Analysis 12 results shown in Table 4.7 are displayed on the monitoring screen (see Fig. 4.1).

We can see that in 2012 the first threshold level was twice exceeded and once (in July) the second threshold level.

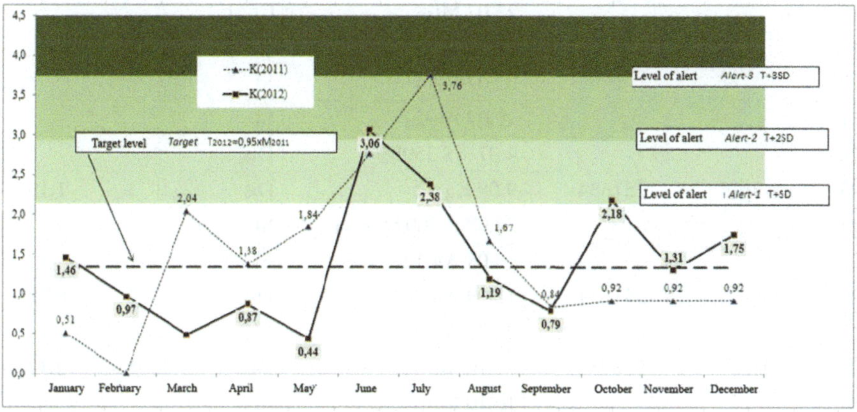

Fig. 4.1 Screen for monthly monitoring of the state indicator of the S level of commercial aviation in the Russian Federation in 2012—(indicator–the total number of ACCIDs and serious incidents per 100,000 h of flight)

Table 4.7 Calculation of the state monthly S indicator for 2012

	Month	n, hour	Event	Type	N	K_S
1	Jan	205882	11.01 Mi-8	SI	3	**1,46**
2			28.01 A-320	SI		
3			31.01 ATR-42	SI		
4	Feb	205882	02.02 An-124-100	SI	2	**0,97**
5			24.02 Yak-18	SI		
6	Mar	205882	21.03 DA-42	SI	1	**0,49**
7	Apr	228758	02.04 ATR-72	Dis	2	**0,87**
8			29.04 Mi-8	SI		
9	May	228758	22.05 Tu-214	SI	1	**0,44**
10	Jun	228758	11.06 Mi-2	Crash	7	**3,06**
11			06.06 Tu-204	SI		
12			05.06 ATR-42	SI		
13			16.06 L-13VVM	Crashe		
14			11.06 An-2	Dis		
15			18.06 An-2	SI		
16			23.06 Boeing 757	SI		
17	Jul	251634	2.07 Mi-2	Crash	6	**2,38**
18			23.07 Mi-8	Crash		
19			26.07 Ka-32	Crash		
20			02.07 R-44	SI		
21			19.07 An-2	SI		
22			4.07 AS 350B3	Dis		
23	Aug	251634	9.08 Ka-32	Dis	3	**1,19**
24			24.08 P-2002 Sierra	SI		
25			13.08 An-124	SI		
26	Sep	251634	12.09 An-28	Dis	2	**0,79**
27			28.09 An-2	SI		
28	Oct	228758	17.10 Mi-8T	Dis	5	**2,19**
29			16.10 Mi-8	SI		
30			25.10 ATR-42	SI		
31			14.10 Mi-171	SI		
32			27.10 Falkon-9000	SI		
33	Nov	228758	17.11 Mi-8	Crash	3	**1,31**
34			21.11 An-26	Crash		
35			05.11 Tu-204	SI		
36	Dec	228758	29.12 Tu-204	Dis	4	**1,75**

(continued)

Table 4.7 (continued)

	Month	n, hour	Event	Type	N	Ks
37			03.12 An-148	SI		
38			18.12 R-44P	SI		
39			20.12 Tu-204	SI		
	Total	2745096			39	

4.4 Methods of the Conventional Safety Performance Indicators Calculation and Monitoring in the Aviation Enterprise

4.4.1 Calculation of the Indicators According to ICAO Recommendations

Following ICAO recommendations, the service providers under the SMS development of indicators of S, their target, and threshold levels.

The government establishes the procedure for the approval and adoption of the set of indicators S provider. This statement of acceptable levels of safety in the SSR relates to the management of risk at the state level, and monitoring and indicators in the SMS of each service provider is part of maintaining S (correct—"confirmation of the level of S"—"safety assurance" [22]).

Indicators and levels of S, the airline must conform to the indicators and levels adopted in the state.

We must recognize that now in Russia:

- no national indicators of S, and, consequently, their target levels and thresholds;
- different opinions of the role and place of the indicators of the level of S, including statistical, in the SMS of service providers;
- no generally accepted methodology for the calculation of the indicator of the level of S in the airline.

The target level of S should be intuitive, easy to count, and check, and this single rule should be based on data about the state of S, who are required to inform the state Supervisory authorities, all airlines. Currently, the list includes in addition to ACCID incidents and industrial accidents. Under certain disadvantages, RIAI-98 [23], the data are considered objective characteristics of the level of S.

As an indicator of S in the airline proposes to use a measure K_{AL} based on the statistics of incidents in the airline by 1 thousand flying hours

$$K_{AL} = \frac{\sum N_{AL}^{IN}}{m} \times 1000, \tag{4.6}$$

where N_{AL}^{IN}—number of incidents in AL for a certain period.

It is proposed to establish acceptable levels of S in the airline using the state levels described above. To do this, it is necessary to establish a correspondence between the acceptable levels of the state indicator of the K_{St} and the indicator K_{AL} through the "compliance coefficient" [9]

$$\eta = \frac{N_{St}^{IN}}{N_{St}^{(ACCID+SI)}}, \tag{4.7}$$

where—N_{St}^{IN} number of incidents in the state;

$N_{St}^{(ACCID+SI)}$ the total number of ACCIDs and serious incidents in the state for a certain period.

The approach to calculating η requires separate consideration.

Often, when estimating the value of η, the "Rule 1: 10: 30: 600" used in the ICAO SMM, 1st ed. 2006. But the pyramid given there was referred in the English text to "industrial safety", that is, to labor safety according to statistics from 1969, the basis is the pyramid proposed by H. Heinrick in 1931, and "the state of safety in the aviation industry" appeared during the translation. According to the text of the SMM, as well as other ICAO guidelines, the terms "*incident*" and "*accident*" do not always correspond to the terms "*incident*" and "ACCID" according to [23] and ICAO Annex 13 "Investigation of Aircraft Accidents and Incidents".

Given different estimates of η and differences in understanding the types of events, it is correct to use the statistics of aviation in the Russian Federation (see Tables 4.8 and 4.9).

The results of calculating the compliance factors η for commercial aviation of the Russian Federation according to the data of the Federal Air Transport Agency for 10 years are summarized in Table 4.10.

Threshold levels for an airline are calculated as

$$K_{AL}^{A} = \frac{K_{St}^{A}}{100}\eta, \tag{4.8}$$

Table 4.8 Statistics of ACCID and incidents in the commercial CA of the Russian Federation

Year	2009	2010	2011	2012	2013	2014	2015	2016	2017	2018	Total
Aviation accidents (ACCID)	12	12	18	14	12	15	12	16	15	19	145
Disasters	6	5	10	7	5	7	5	6	7	9	67
Crashes	6	7	8	7	7	8	7	10	8	10	78
All incidents	838	925	806	868	911	887	778	812	878	840	8543
Serious incidents (SI)	16	33	23	23	26	34	36	26	22	31	270
Incidents	822	892	783	845	885	853	742	786	856	809	8273
ACCID + SI	28	45	41	37	38	49	48	42	37	50	415

Table 4.9 Actual AE ratios in a commercial CA of the Russian Federation (actual "D: C: SI: I" pyramids for 10 years, compare with 1: 10: 30: 600)

Year	2009	2010	2011	2012	2013	2014	2015	2016	2017	2018	Total
Disasters	1	1	1	1	1	1	1	1	1	1	1
Crashes	1	1	1	1	1	1	1	2	1	1	1
SI	3	7	2	3	5	5	7	4	3	3	4
I	137	178	78	121	177	122	148	131	122	90	123

Table 4.10 Estimated values of compliance factors

Year	2009	2010	2011	2012	2013	2014	2015	2016	2017	2018	10 years
Coeff. η	29,9	20,6	19,7	23,5	24,0	18,1	16,2	19,3	23,7	16,8	20,6

For example, using the results of clause 4.3 with a compliance coefficient for 2011 $\eta = 19.7$ calculation by the formula (4.8) gives the following threshold levels:

- threshold level 1—"yellow"—0, 43;
- threshold level 2—"red"—0.58.

The airline develops indicators of less significant events and their threshold levels for "domestic consumption" according to the same ICAO recommendations similar to the method for calculating the state indicator and levels given in clause 4.3 based on the airline's actual results, taking into account the planned increase in S level.

4.4.2 Monitoring of Conventional Safety Performance Indicators

One of the possible calculations and monitoring options is presented. "Internal indicators" based on the total risk level may also be used. The "internal indicator" must ensure the speed of assessment.

It is proposed that taking into account the ICAO approach, the current level of S in the AL also be calculated based on a moving average.

The moving average is widely used to align time series in which there are sharp fluctuations, not only in the economy but also in various fields of technology, including statistics of aviation authorities, equipment manufacturers, and airlines. Examples: UK document CAP-681 [17]—(5-year smoothing period), Airbus PLG program (3-month period), annual S Airline analysis of British Airways (3-month period).

The calculation of the number of events per 1000 flights with a moving average (period of 52 weeks—1 year) is carried out according to the formula

$$K = \frac{\sum_{i=1}^{52} Ki}{52}, \tag{4.9}$$

where $K_i = \frac{N_i}{n_i} \times 1000$—weekly S for the i-th week;

N_i—number of events taken into account for the i-th week;

n_i—number of flights for the i-th week.

In [24], this monitoring method was compared with two other options: a simple weekly calculation and a calculation based on the increase since the beginning of the year. It is shown that a simple calculation of the weekly indicator is unacceptable (spasmodic chart). Monitoring according to the results of the calculation on an accrual basis distorts the dynamics of the indicator.

The method is implemented at Volga-Dnepr Airlines and AirBridge Cargo.

Integral indicator monitoring.

The calculation of the integral indicator of the S level is described in para 4.1. Siberia Airlines and Globus Airlines have implemented a method of weekly monitoring of two integrated S indicators: the incident rate and the general indicator, calculated according to formula (4.1).

When calculating the indicator of incidents, incidents are taken into account, and when calculating the total indicator, all events are taken into account. Monitoring is carried out on a moving average with a smoothing period of 13 weeks (see Fig. 4.2).

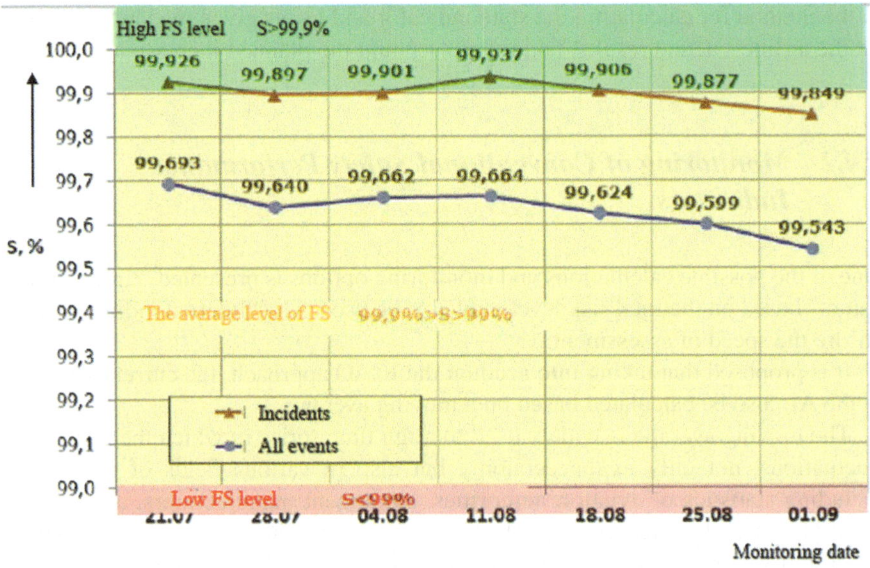

Fig. 4.2 S integral indicator monitoring chart

4.5 Methods of the New Safety Performance Indicators Calculation and Monitoring in the Aviation Enterprise

4.5.1 The Rationale for Two Criteria of Safety Assessment. Monitoring of the Risk R Indicator "Risk Ratio for Deviations and Events" by the Method of the Cumulative Sum

The airline monitoring the indicators in one degree or another is used to inform decisions and practice of the "traffic light" model. If the indicator reaches the "red zone", the airline requires the adoption of measures for its reduction. There are two problems.

1. Statistical indicators or do not take into account the difference in the severity of events or consider it not fully. Such a formal approach is not sufficient for the internal index for developing solutions and plans.

Attempts to take into account the severity of an event are usually reduced to the fact that the airline for the calculation of the internal rate of S includes not only incidents but also, with the reduction coefficient, less significant events ("harbingers").

One of the variants of classification of events is given in [25] (see Sect. 2.1).

One of the variants of differentiation events is implemented in the SMS airline "Aviashelf" [8]. Another option is the use considered an integral indicator. The disadvantages of these calculation methods are noted in Sect. 4. 1.

2. All classification methods divide the events based on formal signs of belonging to one of the groups (incidents, severe incidents, and precursors), i.e., "one-dimensional" scale. In practice, when planning inevitably takes into account two features: actual damage and potential consequences. For example, in dangerous proximity of aircraft in the air, according to [23, 26] is a severe incident. The potential danger is extremely high; this is a potential disaster, but the actual material damage.

In another case, engine failure, requiring a write-off for aircraft with four engines, nothing more than a complication of flight conditions.

Before decision maker inevitably arises the question: how to allocate resources between activities? This task can be considered as a two-criterion task of decision-making. For the two indicators and the author developed the method of factor analysis of parameters of S.

The following indicators are proposed as S indicators:

- RRDE coefficient;
- event damage indicator EDI in monetary terms.

The calculation of the EDI is obvious: the appropriate airline service can easily calculate the actual damage from events.

This section presents the results of a study on the application of the cumulative sum method for monitoring RRDE, and in para 4.5.2 a new decision-support method for S is presented, based on a factor analysis of the proposed indicators of RRDE and the EDI.

The cumulative sum method is developed based on the Shekhar control charts., Which is a graphical description of the state of the technological process? The description of the method is borrowed from the textbook Prof. Orlov [27]. The parameter estimates the normal course of the process. The regulation limits are established that limit the range of permissible values. The exit of the point beyond the control border serves as a signal of the process being disordered.

The method of Shekhart's control charts is used to control technological processes, but it takes into account only the current values of the controlled parameter and does not take into account its previous behavior.

More suitable for the tasks of assessing changes in the level of S is the sum method, based on the accounting of cumulative amounts.

It is proposed to use this method to control the monthly and weekly RRDE. The cumulative amount of K_r is calculated as

$$K_r = \sum_{g=1}^{r} (\text{RRDE}_g - M) = K_{r-1} + (\text{RRDE}_r - M). \tag{4.10}$$

$$g = 1, 2 \ldots .r.$$

where M—the sample average over a period during which the monitored parameter (in this case, RRDE) was normal.

The value of M can be adjusted based on an expert survey. As an example, consider the calculation based on the data of one of the Russian airlines for 2012 Table 4.11 shows the total monthly estimates of RRDE.

Parameter M is calculated as the average RRDE for 6 months of 2012: M = (559 + 253 + 489 + 152 + 280 + 36)/6 = 295.

Cumulative amounts are calculated for the subsequent period, in this case, for the next 5 months, according to the formula (4.10).

For the month of the beginning of monitoring in July, S = 126−M = 126−295 = −169.

For the next month–August: S = −169 + (463−295) = −1, etc.

The results of the calculation of cumulative amounts by the formula (4.10) for 5 months are shown in Table 4.12, and the graph in Fig. 4.3.

The graph shows the change in the level of S relatively acceptable. The method works better with large volumes and frequent observations (see Fig. 4.4).

The graph shows that from the middle of August to September, there was a deterioration in the state of S in the airline. A measure of quantification of deterioration is the slope of the graph curve, which can be measured according to GOST [26].

Calculation methods and tools will require adaptation to the needs of the airline.

Table 4.11 RRDE monthly values

Month	Jan	Feb	Mar	Apr	May	Jun	Jul	Aug	Sep	Oct	Nov
RRDE	559	253	489	152	280	36	126	463	359	140	274

Table 4.12 Calculation of cumulative amounts for monthly monitoring

g (month)	1 (Jul)	2 (Aug)	3 (Sep)	4 (Oct)	5 (Nov)
RRDE g-M	−169	168	64	−155	−21
Sr	−169	−1	63	−91	−112

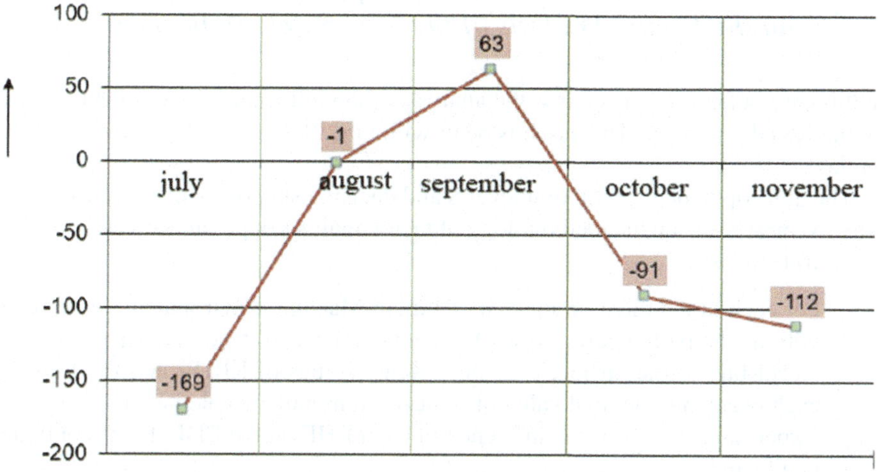

Fig. 4.3 Map of cumulative amounts of RRDE for 5 months of 2012

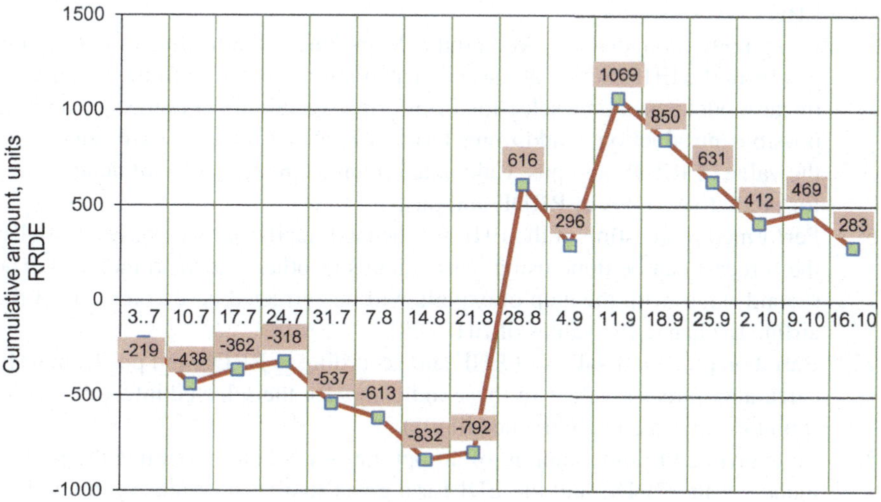

Fig. 4.4 RRDE card with weekly monitoring from 07/03 to 10/16/2012

The proposed indicator and monitoring method are free from the above disadvantages and can be used in the SMS of an airline.

4.5.2 Method for Decision-Making Support Based on the Factor Analysis of the New Safety Indicators

In this case, a factorial analysis is the analysis of the influence of operational factors on the RRDE and EDI. This issue is dealt with in [29]; here is a brief review of the method.

The development of prediction level S and decision support with measuring their effectiveness based on the methodology of factor analysis of parameters of S consists of several stages:

1. For each of the three groups, HF "Human-Machine-Environment" is selected with the help of experts a few of the enlarged hazard factors (EHF).
2. According to the airline, over three years values of RRDE and the EDI for each event and the total value of indicator a month are calculated.
3. Expert assessment of the influence of each EHF on the EDI of the month are conducted.
4. Forming the initial set of statistics, 36 vectors, and values of EHF, RRDE, and EDI for 36 months. The dimension of the vectors depends on the number of EHF.
5. Using regression analysis, we find the dependence of indicators of RRDE and pus from the EHF. Series of variable calculations comes from the underlying linear model based on a selection of informative subsets of regressors and the possible introduction of additional terms [27]. As a result, the EHF can predict the value of RRDE and pus. Calculated point estimates and confidence limits for predictive values of RRDE and pus.
6. Performed forecasting of the EHF for the next quarter in two phases. Initially, the forecast can be done using least squares or other statistical methods. The second stage, with the help of experts will be corrected based on plans of the airline to change the values of EHF [28].
7. Based on predicted values of EHF and according to RRDE and pus from EHF calculated forecast values of these indicators for the selected interval; predict (point estimates and confidence limits).
8. Let's proceed to the planning of an optimal set MD to maximize the performance of the RRDE and the EDI for a given total cost of interventions. It is necessary to form the definition of economic units ("resources aimed at S", "the effect of an intervention S", "related costs", "damages direct, indirect", "resource efficiency".
9. We are developing methods of calculation used economic variables based on data in accounting and managerial accounting, analyze the existing airline

system costs S and opportunities for improvement, develop methods of evaluation-related costs.

10. We are developing a list of MDs used by the airline and a method for calculating the effectiveness of the implemented MDs taking into account cost estimates, a mathematical model for predicting the effectiveness and choosing priority MDs for HF according to the impact on the federal budget, and methods and techniques for optimizing the selection of MDs.
11. We are developing a methodology for testing the method, we are testing the methodology for factor analysis of S indicators and the methodology for choosing events in the airline, and in particular, we calculate the forecast for the next year and compare the results with actual data.

Used conditional, but close to real data on events in one of the airlines for the year and expert estimates of qualified specialists. A simplified version has been adopted—6 enlarged hazard factors (EHF):

HF Crew Errors;
MN Errors of engineering personnel during maintenance;
PP Failures of the engine and its systems;
AC Failure of aircraft systems;
EN An environmental impact;
EH The impact of the artificial environment

For each event during the year, the RRDE value is calculated (according to the airline). The influence (contribution) of each of the EHF to the deterioration of the state of S for a month is expertly determined on a 10-point scale (0-no impact, 10-maximum impact). The evaluation results are given in Table 4.13.

2. According to the data in Table 4.13, source data tables are generated for input into the STATISTICA program (Tables 4.14 and 4.15).

Based on these tables, the STATISTICA program creates a basic linear regression model of the dependence of indicators on factors:

$$P = \beta_0 + \beta_1 F_1 + \cdots + \beta_i F_i + \cdots + \beta_n F_n, \quad (4.11)$$

hier P–RRDE or EDI, dependent variables of the regression model;

F_i–enhanced hazard factors (EHF)-independent variables.

The model allows you to predict the performance of RRDE and the EDI by the values of EHF (Fig. 4.5).

Based on the results of recent years, measures are proposed (managerial decisions-MD) to increase the level of S. Each MD is evaluated by experts. Experts are invited to evaluate how much % the impact of each EHS on S will decrease with the implementation of each of the MD. For example, Table 4.16 shows MDs for improving the reliability of aircraft, training for engineering and technical personnel, and maintaining airworthiness, and Kij is calculated for reducing the j-th MD of the influence of the i-th EHF on the S.

Table 4.13 RRDE and EDI safety indicators and the results of expert assessments of the impact of EHF

N	Date	A place	Event circumstances	Event reason	RRDE Matrix		Indicators		Enhanced Hazards					
					Result	Barrier	RRDE	EDI, $	FO	MN	PP	AC	EN	EH
1	11.1	Burgas norf	Increased vibration of the 2nd MDU, breakage of 2 blades of the turboprop	Design and production disadvantage	2	2	20	1200000						
January							**20**	January	**0**	**0**	**3**	**0**	**0**	**0**
2	11.2	Novos. Irkuts	Destruction of the inner leaf of the compartment of the front landing gear	Fatigue crack	1	1	1	0						
3	14.2	Leipz	Rough landing. Landing below a minimum	Violations of SOP, crew	5	3	500	0						
February							**501**	February	**8**	**0**	**0**	**3**	**0**	**0**
4	14.3	Prestwick	Surge when turning on the reverse, turning off the 2nd dv crew	The destruction of the pipeline 18T.08.14.140	3	3	100	2000						
5	15.3	Bourgas	The harvest of the first row of the main landing gear	Destruction of main landing gear constructs	2	3	20	20000						
6	18.3	Ulyanovsk	Scoreboard “TAKE OFF FORBID.” Take off is stopped	Bit wire 607 F5-1	3	2	20	0						

(continued)

Table 4.13 (continued)

N	Date	A place	Event circumstances	Event reason	RRDE Matrix		Indicators		Enhanced Hazards					
					Result	Barrier	RRDE	EDI, $	FO	MN	PP	AC	EN	EH
March							**140**	March	**0**	**1**	**4**	**4**	**0**	**0**
7	10.4	Muscat	Contact of the bird in the 6th section of the slat.		2	2	4	0						
April							**4**	April	**1**	**0**	**0**	**0**	**2**	**0**
8	1.5	Kandahar	Hard landing.	Crew error	2	3	20	0						
9	5.5	Ahmedabad	Actuation of fire extinguishing phases 1 and 2	Design and production disadvantage	3	3	100	261000						
10	6.5	Kandag Muscat	Getting into the turbulence zone	Unjustified forecast	1	1	1	0						
11	14.5	Bagram	Interrupted take-off, surge of the 4th engine	Design and production disadvantage	3	3	100	300008						
12	23.5	Kuwait	The 1st fire extinguishing stage of MDU No. 2 worked	Design and production disadvantage	2	3	20	200000						
13	25.5	Dakar	Harvesting of the 5th row of the main landing gear of the left and right struts	Error during maintenance	2	2	4	20000						
14	28.5	Canoe - Honolu	Harvesting of the 5th row of the main landing gear of the left and right struts	Error during maintenance	2	2	4	20000						

(continued)

Table 4.13 (continued)

N	Date	A place	Event circumstances	Event reason	RRDE Matrix		Indicators		Enhanced Hazards					
					Result	Barrier	RRDE	EDI, $	FO	MN	PP	AC	EN	EH
May							**249**	May	**2**	**1**	**3**	**6**	**2**	**0**
15	18.6	Ashgabat	Non-localized destruction of the D-18T engine	Design and production disadvantage	3	3	100	1650000						
16	27.6	Ashgabat	Surge after installing Engine control lever on flight small gas. Off 2nd door	Not opening KSD valves.	3	3	100	300002						
June							**200**	June	**0**	**1**	**5**	**0**	**0**	**0**
17	7.7	Kuwait	A bird hit. Without consequences	Environment	1	1	1	0						
18	8.7	Dushanbe	Failure of the 4th engine on the run-up	Design and production disadvantage	3	2	20	300009						
July							**21**	July	**0**	**0**	**3**	**0**	**2**	**1**
19	10.8	Irkutsk-2	Surge, self-shutdown of the 1st engine on the take-off run	Design and production disadvantage	3	3	100	300006						
20	17.8	St. Petersburg	Non-closing valves KPV-KVD on 4 engine	Design and production disadvantage	2	1	2	300005						
21	27.8	Baku	The fire extinguishing system	Alarm Failure	2	3	20	200000						

(continued)

Table 4.13 (continued)

N	Date	A place	Event circumstances	Event reason	RRDE Matrix		Indicators		Enhanced Hazards					
					Result	Barrier	RRDE	EDI, $	FO	MN	PP	AC	EN	EH
August							**122**	August	**0**	**0**	**4**	**2**	**0**	**0**
22	9.9	Turin	On the take-off surge of the 4th 1st engine	Design and production disadvantage	4	2	500	611058						
September							**500**	September	**0**	**0**	**9**	**0**	**0**	**0**
October							**0**	October	**0**	**0**	**0**	**0**	**0**	**0**
November							**0**	November	**0**	**0**	**0**	**0**	**0**	**0**
23	6.12	Burgas Istres	Surge, self-shutdown of the 2nd engine	Not installed	3	3	100	359000						
24	15.12	Mazar Sharif	Srab Signal. “Fire-2nd engine” off two crew	Damage to the electrical harness	3	2	20	200000						
December							**120**	**559000**	**0**	**1**	**6**	**0**	**0**	**0**
Total								**6243088**						
Average contribution ratios for the Mio model							156	520257	**0,9**	**0,3**	**3,1**	**1,3**	**0,5**	**0,1**

Table 4.14 Input data for the formation of the basic RRDE model

Data: ФА_КСС.sta* (7v by 12c)

	1 RDDE	2 FO	3 MN	4 PP	5 AC	6 EN	7 EH
January январь	20	0	0	3	0	0	0
February февраль	501	8	0	0	3	0	0
March март	140	0	1	4	4	0	0
April апрель	4	1	0	0	0	2	0
May май	249	2	1	3	6	2	0
June июнь	200	0	1	5	0	0	0
July июль	21	0	0	3	0	2	1
August август	122	0	0	4	2	0	0
September сентябрь	500	0	1	9	0	0	0
October октябрь	0	0	0	0	0	0	0
November ноябрь	0	0	0	0	0	0	0
December декабрь	120	0	1	6	0	0	0

Table 4.15 Initial data for the formation of the base model of ED

Data: ФА_ПУС.sta* (7v by 12c)

	1 EDI	2 FO	3 MN	4 PP	5 AC	6 EN	7 EH
January январь	1200	0	0	3	0	0	0
February февраль	0	8	0	0	3	0	0
March март	22	0	1	4	4	0	0
April апрель	0	1	0	0	0	2	0
May май	801	2	1	3	6	2	0
June июнь	1950	0	1	5	0	0	2
July июль	300	0	0	3	0	2	1
August август	800	0	0	4	2	0	0
September сентябрь	611	0	0	9	0	0	0
October октябрь	0	0	0	0	0	0	0
November гняюрь	0	0	0	0	0	0	0
December декабрь	559	0	1	6	0	0	1

It can be seen that these MDs, although related to the operation of the EAS and the reliability of equipment, are aimed at different goals and are not alternatives. Therefore, it does not make sense to build the Edgeworth—Pareto set in the coordinates of RRDE-EDI. Based on the obtained regression models, it is possible to evaluate the expected effectiveness of MD for priority selection.

As is known, the multicriteria decision-making problem can either be reduced to a single-criterion one and solved by the linear programming method [30], or solved using "human–machine procedures" using instructions from the work of academician O.I. Larichev [31].

Variable	Predicting Values for (ФА_К(КРОС variable: КС КРОС B-Weight	Value	B-Weight * Value
FO	69,3694	0,900000	62,4325
MN	2,7106	0,300000	0,8132
PP	48,9202	3,100000	151,6527
AC	0,3950	1,300000	0,5135
EN	0,0893	0,500000	0,0446
EH	-72,2860	0,100000	-7,2286
Intercept			-53,6533
Predicted			154,5746
-95,0%CL			82,5072
+95,0%CL			226,6420

Variable	Predicting Values for (ФА_ПУС.s variable: ПУС B-Weight	Value	B-Weight * Value
FO	-66,256	0,900000	-59,630
MN	-478,173	0,300000	-143,452
PP	57,448	3,100000	178,090
AC	158,075	1,300000	205,498
EN	-136,033	0,500000	-68,016
EH	779,872	0,100000	77,987
Intercept			173,708
Predicted			364,184
-95,0%CL			-144,232
+95,0%CL			872,601

Fig. 4.5 Forecast of RRDE and EDI on the monthly average impact of EHF

The sequence of calculations is as follows:

1. Based on the calculations of Table 4.15 (coefficients Mio of the contribution of the i-th EHF in the model without taking into account MD) and Table 4.16 (coefficients K_{ij} of the influence of the jth MD in the i-th EHF), new coefficients of the contribution of the i-th EHF in the M_{ij} model are calculated

$$M_{ij} = M_{i0} K_{ij} \tag{4.12}$$

The calculation results according to the formula (4.12) are summarized in Table 4.17.

2. RRDE and EDI are predicted on the constructed regression model in the STATISTICA program, taking into account new coefficients for each MD.

3. The changes in the RRDE and the EDI are calculated provided that each j-th MD is implemented

$$\Delta RRDE_j = RRDE_o - RRDE_j; \ \Delta EDI_j = EDI_o - EDI_j; \tag{4.13}$$

4. In this case, the criteria for minimizing costs will inevitably be added to the two criteria used in the models (RRDE and EDI) in practice. Therefore, to simplify the problem, we introduce two criteria for the effectiveness of MD

$$C_1^j = \frac{\Delta RRDE_j}{Q_j}, \ C_2^j = \frac{\Delta EDI_j}{Q_j} \tag{4.14}$$

where Q_j—the cost of j-th MD.

We calculate the values of the criteria for all MDs using the formula (4.12).

Table 4.16 Expert assessments of the impact of MD on the EHF

Title of Management Decision (MD)	Cost of MD, $	Coefficients Kij					
		FO	MN	AC	PP	EN	EH
1. Conduct continuing education courses	60000	0,000	0,035	0,015	0,009	0,004	0,005
2. To purchase the control panel the control panel of the electronic control system 18 (TIOESU)	16400	0,000	0,000	0,005	0,020	0,005	0,000
3. To purchase a stand for repair and control of products made of polymer-composite material	26230	0,000	0,000	0,018	0,020	0,000	0,000
4. Training engineering and technical personnel for the diagnosis of engine. in the state enterprise "Ivchenko"—10 people	42000	0,000	0,035	0,008	0,005	0,013	0,005
5. Training engineering and technical personnel in repairing aircraft and electronic equipment in the Saratov Electrical Appliance Plant—6 people, NPO Nauka—2 people	16000	0,000	0,035	0,005	0,003	0,006	0,003
6. Buy a washing unit D-18	300000	0,000	0,008	0,000	0,078	0,003	0,000
7. Buy AS-Pro for battery control	12700	0,000	0,005	0,018	0,010	0,000	0,000
8. Buy a stand P-514 for servicing the blocks of equipment for internal communication and switching to the central laboratory of instrument and radio equipment	201639	0,000	0,010	0,053	0,013	0,000	0,000
9. Monitoring fuel control. D-18T equipment in flight	15000	0,005	0,005	0,000	0,010	0,000	0,000

5. We turn to normalized values.

For criterion C_1, the normalized value C_1^{j*} is calculated as

$$C_1^{j*} = \frac{C_1^j - \min C_1}{\max C_1 - \min C_1},$$

where $\max C_1$ and $\min C_1$ are the maximum and minimum value of the criterion.

For criterion C_2 is a similar calculation.

Table 4.17 Coefficients of the contribution of EHF to the regression model, taking into account the predicted effect of the implementation of MD

MD	Coefficients of M_{ij}					
	FO	MN	PP	AC	EN	EH
1	0,900	0,290	3,073	1,281	0,498	0,100
2	0,900	0,300	3,038	1,294	0,498	0,100
3	0,900	0,300	3,038	1,277	0,500	0,100
4	0,900	0,290	3,085	1,290	0,494	0,100
5	0,900	0,290	3,092	1,294	0,497	0,100
6	0,900	0,298	2,860	1,300	0,499	0,100
7	0,900	0,299	3,069	1,277	0,500	0,100
8	0,900	0,297	3,061	1,232	0,500	0,100
9	0,896	0,299	3,069	1,300	0,500	0,100

6. We use the simple human-machine procedure from [31], based on the introduction of weights w_1 and w_2 of the importance of the criterion, in which the general criterion

$$C = C_1 w_1 + C_2 w_2, \ w_1 + w_2 = 1. \tag{4.15}$$

We assume that the decision maker can establish the values of these weights depending on how he assesses the correlation of the importance of preventing relatively small real damage from current events and the potential danger of a serious aviation event with major damage. Suppose that the following values are set in this airline: $w1$ =0.6; $w2 = 0.4$.

We calculate the values of the general criterion according to the formula (4.15) and sort the MD by decreasing C.

The results of steps 1–6 are summarized in Table 4.18.

Table 4.18 Assessment of the effectiveness of MD by the general criterion

MDj	EDIj	ΔEDI	C_2^j	C_2^{j*}	RRDEj	ΔRRDE	C_1^j	C_1^{j*}	C^j
2	867822	4779	0,291	1,000	222,8	3,89	0,00024	1,000	**1,000**
7	869342	3259	0,257	0,874	224,5	2,13	0,00017	0,738	**0,792**
3	867691	4910	0,187	0,622	222,6	4,06	0,00015	0,689	**0,662**
9	870130	2471	0,165	0,541	224,6	2,09	0,00014	0,630	**0,594**
6	855459	17142	0,057	0,150	212,6	14,09	0,00005	0,281	**0,229**
1	869541	3060	0,051	0,128	225,7	0,90	0,00001	0,161	**0,148**
4	870693	1908	0,045	0,108	226,6	0,05	0,00000	0,108	**0,108**
8	869433	3168	0,016	0,000	223,8	2,80	0,00001	0,156	**0,094**
5	871331	1270	0,079	0,231	227,1	−0,44	−0,00003	0,000	**0,092**

Table 4.19 Ranking MD according to the degree of decrease in predicted efficiency

п/п	Management decision	Cost	Single criterion
1	To purchase a control panel, a control panel for an electronic control system—18 (TIOESU)	16400	1,000
2	Purchase an AS-Pro for battery monitoring	12700	0,792
3	Purchase a stand for repair and control of products made of polymer-composite material	26230	0,662
4	Monitoring fuel control parameters. D-18T equipment in flight	15000	0,594
5	Purchase a gas-air duct flushing unit for the D-18 engine	300000	0,229
6	Conduct continuing education courses	60000	0,148
7	Training in the diagnosis of the engine in the state enterprise "Ivchenko"—10 h	42000	0,108
8	Purchase stand P-514 for servicing and repairing units of internal communication and switching equipment at the central laboratory of instrument and radio equipment	201639	0,094
9	5. Training in the repair of aviation and electronic equipment in the Saratov Electrical Instrument Plant named after Ordzhonikidze	15000	0,092

The decision maker can use the result in determining the priority of the implementation of the proposed management decisions (MD) (Table 4.19).

References

1. Annex 19 to the Convention on international civil aviation. The management of safety. ICAO 2013, p 44
2. Statistical methods analysis of the safety of complex technical systems: Textbook/L. N. Alexander and others–M.: Logos, pp 200–231
3. Sharov VD (2011) Methodology of application of the combined method FMEA-FTA for risk analysis of aviation events. In: Sharov VD, Makarov VP (eds) Scientific bulletin of MSTUCA. No. 174, pp 18–24
4. Shviryaev YV (1992) Probabilistic analysis of the safety of nuclear stations. Methods of execution. YV Shviryaev and others. M: IAE them. IV Kurchatov, p 266
5. 23748-88 GOST B, GOST In 20570-88, GOST In 20436-88 flight Safety, reliability, testability, and repair processability. Goskomstat SSSR, 1988, p 10
6. Zubkov BV (2009) Safety management. Part I. Program of safety. Guide safety management: a training manual. BV Zubkov, PM Polyakov, MV Karmysov. M: MSTU CA, p 132
7. Aviation rules (2004) Part 25. Norms of airworthiness of transport category airplanes, 2nd edn. MAC, JSC "AVIAIZDAT", p 240
8. The materials of the seminar "SMS in civil aviation and preparation for the validation of the ICAO USOAP". [Electronic resource]. M.: air transport, 2008. - 1 electron.opt. Disk (CD-ROM)
9. Zubkov BV (2010) Theory and practice of risk management in aviation enterprises in the development of system safety management. BV Zubkov, VD Sharov. M.: MGTU CA, p 196
10. Plotnikov NI (2008) The Foundation of the methods of calculating the cost of safety.N. And. Carpenters. Scientific Bulletin of MSTUCA, No. 135, pp 67–72

11. Rukhlinskiy VM (2008) A new criterion to quantify the level of safety. Rukhlinskiy VM. Scientific Bulletin of MSTUCA. № 135(11), pp 202–204
12. Zhivetin VB (2009) The System of civil aviation safety (analysis, forecasting, management), vol 17. VB Zhivetin. Institute of problems of risk, Information and publishing center "Bon Anni", Moscow, p 370
13. Analyses the safety status of flights in civil aviation of the Russian Federation in 2003–2013 GS HA fsnst, FATA
14. Safety Report, 2013. IATA, Montreal-Geneva, 2014
15. ICAO 2012 Safety Report, Montreal, 2013
16. NTSB USA [Electronic resource] access Mode. http://www.ntsb.gov/data/aviation_stats.html
17. CAP 681 Global Safety Oversight Review. 1980–1996. CAA UK, 1996
18. Annual Safety Report 2011, EUROCONTROL, February 2012
19. SAFA Ramp Inspection's Guidance Material. Ver. 2, 0, EASA, July 2012
20. Sharov VD (2013) Prediction, and prevention of rolling out of the aircraft outside of the runway. Sharov VD (ed) Lambert, p 112
21. Guide to safety management (SMM) (2013) Doc.9859. ICAO, 3rd edn, p 300
22. Sharov VD (2013) Some problems understanding terms related to the management of safety. In: Sharov VD (ed) Problems of safety in 2013. No. 9, pp 12–21
23. Rules of investigation of aviation incidents and incidents with civil aircraft in the Russian Federation–PRAPI–98. Approved. The RF Government decree of 18 June 1998, No 609. M: Aviaizdat, 1998, p 140
24. Sharov VD (2014) Development of an indicator of safety based on the recommendations of the ICAO. In: Sharov VD (ed) Scientific Bulletin of MSTUCA, No 204, pp 97–104
25. Zubkov BV (2011) Methods of risk assessment in system safety management. In: Zubkov BV, Prozorov SE (eds) Scientific Bulletin of MSTUCA, No 174, pp 7–11
26. Malygin VB (2011) Safety in ATC: a training manual. In: Malygin VB, Nechaev EE (eds) M.: MGTU CA, p 88
27. Orlov AI (2012) Organizational and economic modeling: a textbook in 3 hours Part 3. Statistical methods of data analysis. AI Orlov. M.: Izd-vo MGTU of, NU Bauman, p 624
28. GOST R 50779.45-2002 Statistical methods. Control charts cumulative sums. Publishing House of Standards, 2002, p 17
29. Sharov VD (2013) Identify deviations in the controlling system (on an example of monitoring the level of safety). In: Sharov VD, Orlov AI (eds) Green Controlling: proceedings of the III International Congress in controlling. Under nauch Falko SG (ed). M.: NP "Association of controllers", pp 277–292
30. Wentzel ES (1972) Research of operations. ES Wentzel. M.: Sov. Radio, p 550
31. Larichev OI (2008) Theory and methods of decision making, as well as Chronicle of events in Magical countries: textbook. In: Larichev OI. –M.: University book. Logos, 392 p

Annex 1: Implementation of the Program "Managerial Decisions Efficiency Estimation"

A.1 Introduction: Formulation of the Problem

Based on the study of the experience of the CAST Group (USA), a method for assessing the effectiveness of managerial decisions (MD), which are proposed to reduce the level of risk, has been developed.

A software product "Assessment of the effectiveness of management decisions" in the MS Excel format was developed to implement the algorithm. The program is an MS Excel book containing several sheets: "Data Entry", "Computational Actions", "Result", etc., and several diagrams.

Annex 1 gives an example of using this tool on the contingent data of one of the airlines.

The task is posed as follows. According to the results of the airline's risk assessment, for example, by the ASPPAP program, for a specified period it was established that an excess of the acceptable level of risk was noted for events of three types with the following risk values in US dollars:

- the collision of a working aircraft with the ground (CFIT)—2500;
- the rollout of aircraft beyond the runway during landing (RE)—4000;
- loss of control of aircraft in flight (LOC)—1200.

It is necessary to offer the decision maker (DM) several management decisions (MD) to reduce risk with a forecast of their effectiveness.

V. D. Sharov et al., *Probabilistic-Statistical Methods for Risk Assessment in Civil Aviation*, Springer Aerospace Technology, https://doi.org/10.1007/978-981-16-0092-0

A.2 The Solution to the Problem

A.2.1 Compilation of a List of Hazard Factors and Expert Assessment of their Impact

Based on a survey of experts, the following list of hazard factors (HF) was obtained:

- a significant flight of the landing zone during landing;
- increased landing speed (touch speed);
- improper operation of reverse engine thrust;
- failure of the aircraft hydraulic system;
- false information about the status of the runway (the coefficient of friction);
- poorly conducted preliminary preparation for the flight;
- Failure of the ground proximity early warning system (GPWS);
- insufficient level of English proficiency by members of the flight crew;
- Inadequate crew resource management (CRM);
- Errors in the perception of indications of horizons;
- loss of speed in flight;
- excess of the permissible values of the roll.

An expert survey is conducted on the basis of which the proportion of the risk of an event attributable to each of the hazard factors (HF) affecting this event is estimated. According to the estimates, the share of risk in US dollars is calculated, which falls on each HF.

The results of estimates and calculations are summarized in Table A.1.

A.2.2 Drawing Up a List of Management Decisions

Based on a survey of experts and managers of airline divisions, the following list of management decisions (MD) has been compiled:

- regular repetition of the Flight operations manual of the airline;
- additional control of aircraft maintenance;
- special classes with flight personnel;
- conducting additional training sessions;
- a one-time additional inspection of aircraft hydraulic units;
- appeal to the administration of destination airports in order to improve the awareness of airline crews about the actual status of the runway and the value of the coefficient of friction.

Table A.1 Results of the assessment of the impact of HF on the risk of an event

High-risk event	Event risk assessment, US dollars	Hazard Factor (HF)	The proportion of the risk of the event attributable to the HF	Assessment of the risk share per hazard factor, US dollars
RE	4000	Touchdown zone	0,12	480
		Touch speed overshoot	0,24	960
		Incorrect operation. Reverse engine thrust	0,04	160
		Aircraft Hydrosystem Failure	0,20	800
		Invalid information about the coefficient of friction	0,40	1600
CFIT	1200	Poor preparation	0,50	600
		GPWS Failure	0,05	60
		Inadequate English	0,15	180
		Inadequate CRM	0,30	360
LOC	2500	Error of perception of horizons	0,10	250
		Loss of speed	0,40	1000
		Increased roll	0,50	1250

A.2.3 Assessment of Expected Effectiveness, Cost: "Level of Implementation" of Management Decisions

The expected effectiveness of the impact of each MD on each of the HFs is estimated based on a survey of experts. Efficiency is expressed as the proportion of the risk of each of the HF, which reduces the negative impact of HF.

The cost of each MD can be determined quite accurately.

Under the "level of implementation" of MD is understood the coefficient $K \leq 1$, which shows the extent to which this MD is implemented in the airline. This may be the share of aircraft from their total number in the airline on which new equipment is installed; part of the flight or technical personnel with whom additional training or simulator training has been carried out; part of the aerodromes, to the administrations of which a written request was sent with a request to improve the information of the airline crews, etc.

Table A.2 Projected efficiency and "implementation level" of MD

Management Decision (MD)	Flight operations manual repeat	Maintenance monitoring	Special training	Additional simulator	Inspection of aggregates	Requests to the airport
MD implementation level	0,8	0,5	0,9	0,75	0,3	0,2
Cost of MD, USD	0	200	100	1000	200	20
	The effectiveness of the impact of MD on HF					
HF	0,05	0	0,15	0,50	0	0
The flight of the landing area	0,05	0	0,15	0,60	0	0
Over speed	0,05	0	0,15	0,60	0	0
Incorrect operation	0	0,60	0,0	0	0,7	0
HS failure	0	0	0	0	0	0,2
Invalid information of the coefficient of friction	0,15	0	0,30	0,40	0	0
Low-quality PP	0	0,60	0		0,7	0
GPWS Failure	0	0	0,35	0,30	0	0
English level	0,15	0	0,30	0,55	0	0
Inadequate CRM	0,10	0	0,10	0,65	0	0
An error of perception AH	0,10	0	0,10	0,65	0	0
Loss of speed	0,10	0	0,10	0,60	0	0

The results of expert evaluations of the effectiveness and the expected "level of implementation", as well as data on the cost of MD, are shown in Table A.2.

A.2.4 Performing Calculations

Calculations are performed using the developed software product "Evaluation of the effectiveness of the MD", which is a few related tables MSExcel, which implements the algorithm.

Data entry is carried out in accordance with the instructions.

Instructions for entering data into the program "Evaluation of the effectiveness of MD".

The main goal is to automate the management actions (actions, improvements in S). Data entry is performed in accordance with the regulations. Note: steps 1–4 can be performed in any order, but the category must match the events. All data is entered only in dark yellow cells.

Step 1. Enter the alleged actions in row 3, one in each cell, starting with the NC sheet Input.

Step 2. Select the events or conditions for which will be evaluated by the actions in column e of the worksheet data Entry.

Step 3. Enter the category for each event in the article and sheet Input. The risk will be divided according to these categories. Each event must be assigned to a category. If there is more than one event, write the category name must be the same.

Step 4. Enter once the name of each category in column A of sheet Result, starting with A7. Use the names of the categories, identical to Step OVER.

Step 4A. Enter a description of the severity of the consequences (0–1) of each event in article D of the worksheet. Enter the data.

Step 4B. Enter the probability of recurrence of each event in the article from the sheet Input. (It is proposed to use the probability equal to 1).

Step 5. Enter the estimated efficiency (from 0 to 1) of each action relative to each event as the probability of preventing each event by starting from cell 37 of the sheet Input.

Step 6. Enter a description of the mass implementation of each action (0–1) as a share of an aircraft (flight crew), for which this action is being implemented, starting in cell C5 sheet a Result.

View results. Reducing the overall risk presented in the worksheet Result in cell 330. Reduction of overall risk due to each action presented in cells K30 to ETSO. The percentage risk reduction due to 20 actions shown in the Action Figure. The percentage reduction in the risk category due to all of the actions shown in the Diagram Category.

The appearance of the main sheets of Excel tables is shown in Figs. A.1, A.2 and A.3.

Based on the results of calculations, the program generates diagrams that are provided to the person making managerial decisions for making decisions on the implementation of measures.

The principal summary diagrams are presented in Figs. A.4, A.5, A.6, A.7 and A.8.

A.3 The Conclusion on the Results of Using the Program

The MD effectiveness assessment program allows us to optimize the search for decision makers of event options and their ranking according to the degree of reduction of the risk of ACCID in the context of limited funding.

TYPE OF AE	The proportion of risk of a given event on the given HF (Rik=Ci*Pik, USD	The part of these events that occurred due to this HF(fo contribution) (Pik)	Risk of this event, (Ri) USD	HF	The probability that the HF will not appear after all actions are implemented, or the "prevented share of HF" (Qi)	Eliminated the risk HF after the implementation of all actions ΔR=Ri*Qi	Residual risk of HF Ri'=Ri-ΔRi after all the action	Share of eliminated HF risk ΔRi/Ri	1	2	3
									Name of the embedded action		
									Repeat FOM	Maintenance monitori	Special lesson
									The share of aircraft (crews, v		
									,800	,500	,900
									The effectiveness of each actic		
Roll out RCZ	480	0,120	4000	The flight of the landing area	,481	230,880	249,120	48,1%	0,05	0,00	0,15
Roll out RCZ	960	0,240	4000	Exceeding the touch speed	,543	521,549	438,451	54,3%	0,05	0,00	0,15
Roll out RCZ	160	0,040	4000	Direction of the operated reve	,543	86,925	73,075	54,3%	0,05	0,00	0,15
Roll out RCZ	800	0,200	4000	HS failure	,496	396,800	403,200	49,6%	0,00	0,60	0,00
Roll out RCZ	1600	0,400	4000	Invalid information coefficient c	,040	64,000	1536,000	4,0%	0,00	0,00	0,00
CFIT	600	0,500	1200	Low-quality PP	,550	330,192	269,808	55,0%	0,15	0,00	0,30
CFIT	60	0,050	1200	GPWS failure	,496	29,760	30,240	49,6%	0,00	0,60	0,00
CFIT	180	0,150	1200	English level	,469	84,443	95,558	46,9%	0,00	0,00	0,35
CFIT	360	0,300	1200	Bad CRM	,623	224,132	135,868	62,3%	0,15	0,00	0,30
LOC	250	0,100	2500	An error of perception AH	,571	142,734	107,266	57,1%	0,10	0,00	0,10
LOC	1000	0,400	2500	Loss of speed	,571	570,935	429,065	57,1%	0,10	0,00	0,10
LOC	1250	0,500	2500	Roll large	,540	674,425	575,575	54,0%	0,10	0,00	0,10

Fig. A.1 Fragment of the table for input of source data and calculation of the probability of HF manifestations

TYPE OF AE	The proportion of risk of a given event on the given HF (Rik=Ci*Pik)	HF	probability that the HF will not appear after all actions are implemented, or the "prevented share of HF" (Qi)	Eliminated the risk HF after the implementation of all actions ΔR=Ri*Qi	1	2	3	4	5	6
					Name of the embedded action					
					Repeat FOM	Maintenance	Special lesson	Additional sim	Inspection	Contacting
					Percentage of aircraft (flights) for which the measure (Kj) is implemented),					
					,800	,500	,900	,750	,400	,200
					Eliminated the risk of this event after implementing each j-th action					
Roll out RCZ	480,000	The flight of the landing area	0,481	230,880	24,000	,000	72,000	240,000	,000	,000
Roll out RCZ	960,000	Exceeding the touch speed	0,543	521,549	48,000	,000	144,000	576,000	,000	,000
Roll out RCZ	160,000	Direction of the operated revers	0,543	86,925	8,000	,000	24,000	96,000	,000	,000
Roll out RCZ	800,000	HS failure	0,496	396,800	,000	480,000	,000	,000	560,000	,000
Roll out RCZ	1600,000	Invalid information coefficient	0,040	64,000	,000	,000	,000	,000	,000	320,000
CFIT	600,000	Low-quality PP	0,550	330,192	90,000	,000	180,000	240,000	,000	,000
CFIT	60,000	GPWS failure	0,496	29,760	,000	36,000	,000	,000	42,000	,000
CFIT	180,000	English level	0,469	84,443	,000	,000	63,000	54,000	,000	,000
CFIT	360,000	Bad CRM	0,623	224,132	54,000	,000	108,000	198,000	,000	,000
LOC	250,000	An error of perception AH	0,571	142,734	25,000	,000	25,000	162,500	,000	,000
LOC	1000,000	Loss of speed	0,571	570,935	100,000	,000	100,000	650,000	,000	,000
LOC	1250,000	Roll large	0,540	674,425	125,000	,000	125,000	750,000	,000	,000
0	0,000	0	0,000	0,000	,000	,000	,000	,000	,000	,000

Fig. A.2 Fragment of the table of calculations of the eliminated risk

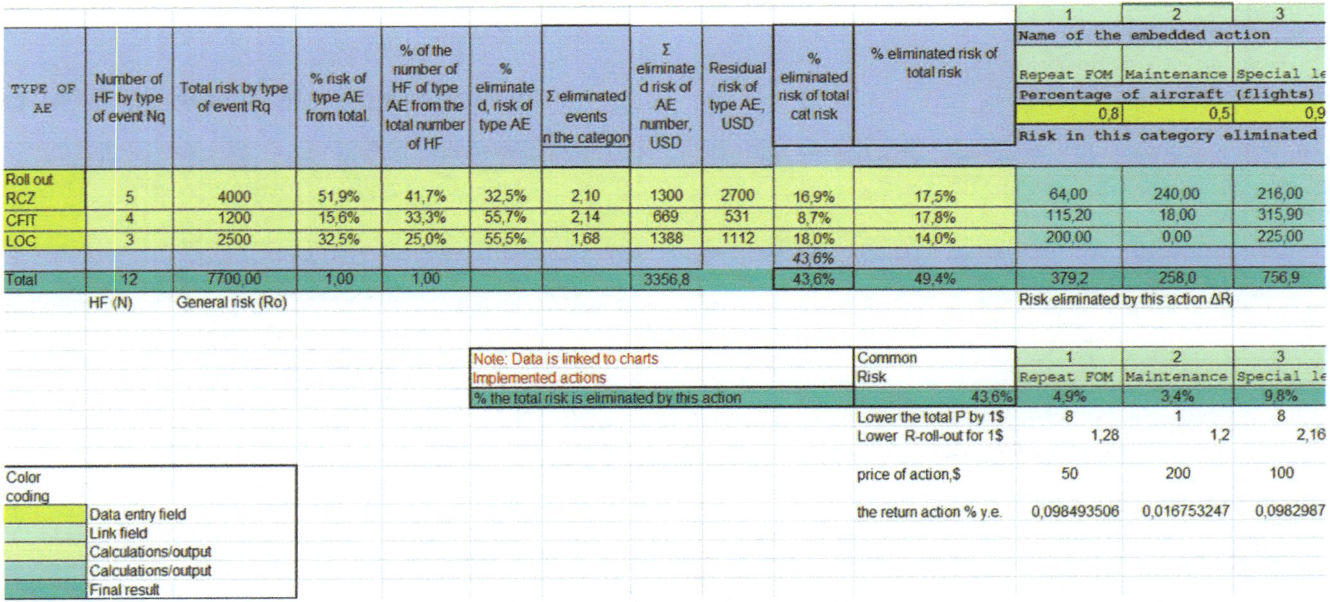

TYPE OF AE	Number of HF by type of event Nq	Total risk by type of event Rq	% risk of type AE from total	% of the number of HF of type AE from the total number of HF	% eliminated, risk of type AE	Σ eliminated events in the category	Σ eliminated risk of AE number, USD	Residual risk of type AE, USD	% eliminated risk of total cat risk	% eliminated risk of total risk	1	2	3
											Name of the embedded action		
											Repeat FOM	Maintenance	Special le
											Percentage of aircraft (flights)		
											0,8	0,5	0,9
											Risk in this category eliminated		
Roll out RCZ	5	4000	51,9%	41,7%	32,5%	2,10	1300	2700	16,9%	17,5%	64,00	240,00	216,00
CFIT	4	1200	15,6%	33,3%	55,7%	2,14	669	531	8,7%	17,8%	115,20	18,00	315,90
LOC	3	2500	32,5%	25,0%	55,5%	1,68	1388	1112	18,0%	14,0%	200,00	0,00	225,00
									43,6%				
Total	12	7700,00	1,00	1,00			3356,8		43,6%	49,4%	379,2	258,0	756,9
	HF (N)	General risk (Ro)									Risk eliminated by this action ΔRj		

		1	2	3
Note: Data is linked to charts	Common			
Implemented actions	Risk	Repeat FOM	Maintenance	Special le
% the total risk is eliminated by this action	43,6%	4,9%	3,4%	9,8%
	Lower the total P by 1$	8	1	8
	Lower R-roll-out for 1$	1,28	1,2	2,16
	price of action,$	50	200	100
	the return action % y.e.	0,098493506	0,016753247	0,0982987

Color coding	
	Data entry field
	Link field
	Calculations/output
	Calculations/output
	Final result

Fig. A.3 Fragment of the table for calculating the eliminated risk and efficiency of MD

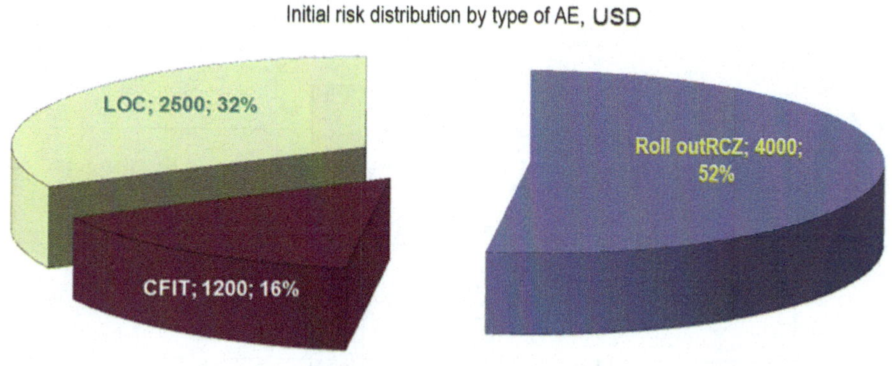

Fig. A.4 Distribution of risk of the event

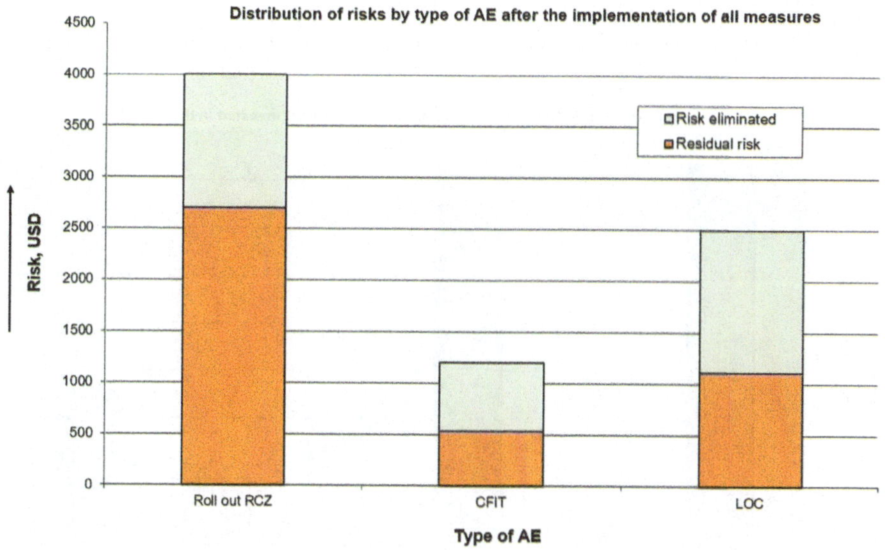

Fig. A.5 The predicted distribution of the eliminated and residual risks in the implementation of all MD events

The program can be easily adapted to the activities of any airline.

The modernization of the method significantly expands its capabilities, making it possible to evaluate the effectiveness of both single corrective actions and their combinations for several indicators, as well as to assess the degree of risk reduction by the categories of aircraft, by the severity of the consequences, causality, types and belonging of aircraft, the territory of their base and application, components of the "Crew-Aircraft-Environment" system, etc.

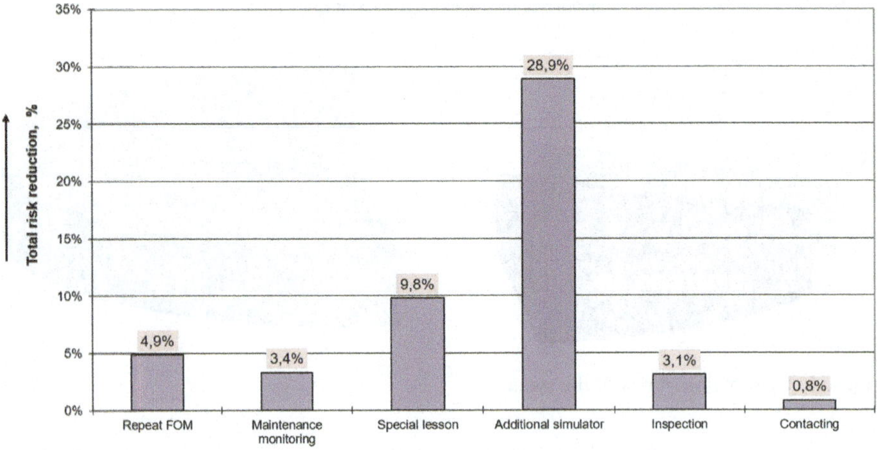

Fig. A.6 The projected reduction in overall risk when implementing each MD

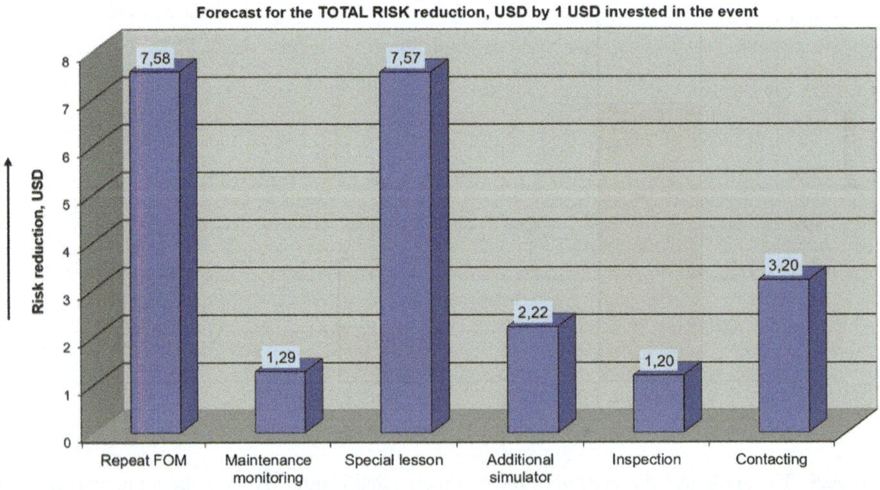

Fig. A.7 Forecast of overall MD productivity

The proposed approach to a priori estimation of the effectiveness of measures, as an element of the implementation of an active anti-aircraft prevention strategy, can be the basis for the algorithmic support of the safety management system, both at the state and the industry (departmental) or corporate levels.

The approach is general and can be used in the development of methodological support for risk management at high accident facilities for various purposes.

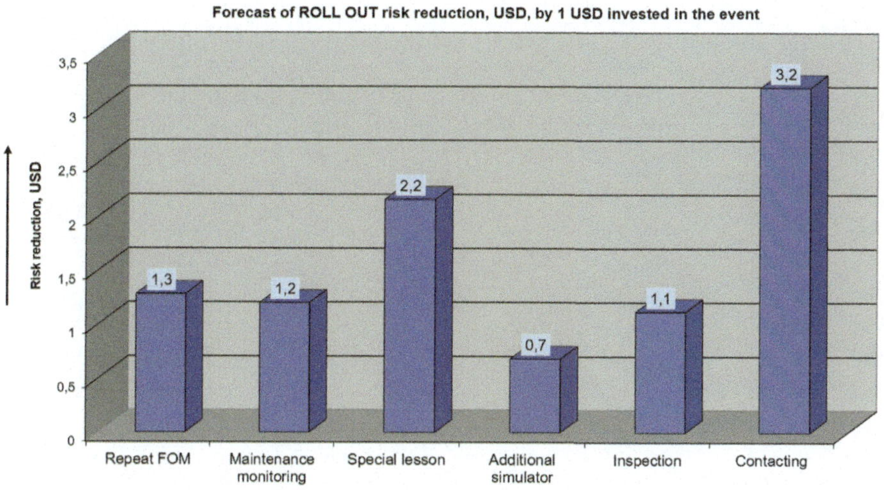

Fig. A.8 The forecast of performance of MD to reduce the risk of rolling out

The manufacturer's authorised representative in the EU is Springer Nature Customer Service Centre GmbH, Europaplatz 3, 69115 Heidelberg, Germany. If you have any concerns regarding our products, please contact ProductSafety@springernature.com

Printed and bound by CPI Group (UK) Ltd, Croydon, CR0 4YY
30/09/2026
02225306-0013